It Was What It Was

A Tale of the 1st Infantry Division in Iraq (2006 - 2008)

Daniel Pace

Grey Zone Ethics LLC

Other works by Daniel Pace:

Grey Zone Ethics

Cover art by Chad Woody

Editing by Randy Surles, Laura Graves, and Anna Simons

Department of Defense Release Statement:

The views expressed in this publication are those of the author and do not necessarily reflect the official policy or position of the Department of Defense or the U.S. government. The public release clearance of this publication by the Department of Defense does not imply Department of Defense endorsement or factual accuracy of the material.

Dedication

Primarily, this book is dedicated to the nine members of the 1st Squadron, 4th U.S. Cavalry who were killed in action during our 2007-2008 rotation to Iraq. May we remember their heroism and sacrifice, and may we all meet again someday to share a drink and a story in Fiddler's Green.

SSG Courtney "Tuba" Hollinsworth; Died September 9, 2007

PFC Christopher M. North; Died April 21, 2007

SPC Braden J. Long; Died August 4, 2007

PFC Michael Pitman; Died June 15, 2007

PFC William Johnson;Died June 12, 2007

SPC Rodney J. Johnson; Died September 4, 2007

PFC Steven Walberg; Died April 15, 2007

SPC Robert J. Dixon; Died May 6, 2007

PFC Aaron Genevie; Died April 16, 2007

This book is also dedicated to our interpreters (especially Ozzie) and to the Iraqi people of *Mahalas* 826, 838, and 840, all of whom risked life and limb to help us rebuild their country. Without their effort and bravery, the above list of slain Americans would have been far longer, and our efforts in Doura would have amounted to far less.

Finally, the book is dedicated to the thousands of families who were shattered irreparably by the GWOT. These broken homes are neither recorded nor commemorated, but that anonymity makes them no less tragic.

May we all remember the costs of war and ensure the desired ends are worth the price we pay to accomplish them.

Acknowledgements

This book would not have been possible without the assistance, feedback, and input of a great many people, including: James Crider, Brett Hamilton, Rich Smith, Mark Ehlers, Robert Ritchie, Gannon Edgy, James Danly, Justin Bakal, Josh Sproul, Aaron McDowell, Miles Blanco, Nathan Banninger, Robert Humphrey, Koky Sisoura, Dave M, Jason Fedish, Lisa Deis, Carly Costello, Chad Woody, Clint Woody, Dan Lamb, John Mclellan, my editors, Randy Surles and Laura Graves (The Story Ninjas), and Anna Simons, *and*, of course, my lovely wife, editor, and lifelong friend, Alycia Pace.

Glossary

9-line (Nin-lin) **n.** a formatted report for calling in either a MEDEVAC or an IED on the radio. The report provides all the information needed to coordinate an air MEDEVAC in as few words as possible.

BIP (Bip) **v. acronym** (Blow-in-place). to destroy something with explosives

Brain Housing Unit (Bran Houz-ing Yoo-nit) **n. slang.** Head. The term is a play on the army's tendency to name everything in a complex and unwieldy manner.

CSH (Kash) **n. acronym** (Combat Surgical Hospital) A major military field hospital. In Baghdad, the CSH we worked with was in the Green Zone, and it wasn't a field hospital at all, as it occupied several large, concrete buildings and had all the emergency medicine capabilities of any U.S. hospital. Anyone requiring more treatment than the CSH could provide was flown back to the even larger and more capable military hospital in Landstuhl, Germany.

Chow Hall (Chow'-hall) **n. slang.** The Army term for a cafeteria. Chow halls vary widely in quality and visual appeal, ranging from Soviet-esque slop lines that specialize in such army classics as SOS (shit on a shingle) to gleaming (usually Air Force) nutritional Meccas.

Combatives (Kum-bat'-ivs) **n.** The Army's hand-to-hand combat instruction program.

Connex (Kon'-ex) **n.** Large, metal, shipping containers designed for commercial and military shipping. Because of their standardized, Lego-like, shapes, and rugged construction, the military repurposes connexes for all sorts of uses, from living quarters, to frozen food storage, to communication centers, to gyms.

COP (Kop) **n. acronym** (Combat Outpost). a small, shitty Forward Operating Base (FOB), whose main redeeming feature is its lack of proximity to everyone in your chain of command you'd rather not encounter on a daily basis.

DHA (Dee'-Haa) **n. acronym** (Detainee Holding Area). A building where units dropped off captured enemy personnel so U.S. military police could screen them and hand them off to the Iraqi judicial system. The DHA we used was in the Green Zone. It was conveniently located near a Green Beans coffee shop, which made taking detainees to the DHA a bit of a field trip for the hot and tired members of our unit.

Duck Hunter (Dukk' hun-ter) **n. slang.** A slang term for air defense artillerymen (ADA), the folks who shoot down enemy aircraft. During the GWOT, when the Army needed infantrymen for counterinsurgency operations, the ADA was heavily poached for personnel and funding.

FOB (Fob) **n. acronym** (Forward operating base). a temporary (but long-term temporary) forward base. Distinguished from a COP or Fire Base by its size, relative permanence, better chow, and more luxurious lavatories. Also, FOBs usually contained the members of your chain of command who you'd rather not encounter on a daily basis.

Fobbit (Fob'-it) **n. slang. Derogatory.** A disparaging term for members of the armed forces who conducted their duties on or around bases. Combat forces accused Fobbits of monopolizing popular, limited resources (such as snacks or popular movies or music), crowding the chow hall, and collecting unearned combat benefits. A YouTube video popularized the term.[1]

Full Battle Rattle (Full-ba-tul-rat'-ul) **n. slang.** The Army term for wearing all the gear necessary to conduct an operation. In Iraq, this comprised a weapon, armor, helmet, eye protection, gloves, knee pads, radio, medical kit, and ammunition. Setups varied a bit by position, but it generally weighed about 50 to 60 lbs.

Full Bird Private (Full-burd'-pry-vet) **n. slang.** An informal term for the specialist rank, which is the highest rank a lower enlisted service member can achieve before becoming a noncommissioned officer. The term is a play off the informal term "full bird colonel," which refers to an army colonel, whose rank is denoted by an eagle. Full Bird Privates are notoriously capable of gaming the army system and shirking responsibility by exploiting their lower-ranking comrades.

GWOT (Jee'-wot) **n. acronym** (Global War on Terror). The collection of military actions undertaken by the United States following the World Trade Center attacks in 2001 and formally ended by President Biden in August, 2021.

HESCO Barrier (Hess'-ko-bear-ee-er). A collapsible metal frame lined with woven cloth. When unfolded and filled with dirt, HESCOs form an explosion and im-

---

1. https://www.youtube.com/watch?v=6Hqv5yBaXaI

pact-resistant brick that is used in FOB construction around the world.

Hooah (who'-uh) **n. slang** A generic army response that can mean anything from "I understand," to "piss off" depending on the speaker's tone, rank, and intent. Also, slang for a private, as in, "Go tell that bunch of young hooahs over there to clean up that trash."

HMMWV (Hum'-vee) **n. acronym** (High mobility, multipurpose, wheeled vehicle). A diesel powered military vehicle originally produced to transport light equipment and personnel around secured areas of the battlefield and to serve in other non-combat functions. During the Global War on Terror, the HMMWV became one of the primary combat vehicles employed by the U.S. military in both Iraq and Afghanistan for most ground operations until it was largely replaced by more capable vehicles.

IED (I'-ee-Dee) **n. acronym** (Improvised explosive device). a homemade bomb.

IVO (In-vi-sin'-it-ee-ov) **adj. acronym** (In-vicinity-of). Near a known object. (e.g. The terrorist is IVO the 4th street mosque).

M249 (Em'-too-for-nine) **n.** The M249 Squad Automatic Weapon, or SAW, is a belt-fed 5.56 caliber light machine gun carried by infantry squads in the U.S. Army. During the GWOT, vehicles often mounted the SAW alongside the MK-19 Automatic grenade launcher, but because of the latter weapon's potential to cause collateral damage, the SAW often became the truck's primary weapon.

M4 (Em-for') **n.** The M4 carbine was the standard weapon of almost all active-duty soldiers during the GWOT. The weapon is 5.56 caliber, which gives it almost no recoil, and carries a 30-round magazine. Accessorizing one's M4 was a favorite pastime of many soldiers in Iraq, and the variety of sights, slings, grips and lights found on M4's across the theater was mind-boggling.

MEDEVAC (Med'-i-vac) **v. acronym** (medical evacuation). the act of transporting a casualty from the point of injury to a medical treatment facility in combat by a dedicated air or ground evacuation asset. It's like calling an ambulance in combat. **N. acronym** (medical evacuation). The act of conducting the above activity (e.g. Joe's hit; call a MEDEVAC!)

Mikes (Miks) **n. slang.** minutes.

Mahala (Moo'-ha-la) **n. Arabic.** A neighborhood. In Baghdad, a person's address contained the *mahala* number, and the *mahala* has some sort of role in the allocation of civil services that was never entirely clear to me.

NVGs (En-vee'-geez) **n. acronym** (Night Vision Goggle). Goggles that allow the wearer to see in the dark by enhancing the ambient light in the environment. During The Surge, most regular army units had PVS-14s, which were monocular devices that left one of the user's eyes unaugmented and gave the other eye the power to see a grainy, green version of reality. PVS-14s aren't thermal sights, so wearing them doesn't give you *Predator*-like white-hot vision, but it *does* allow you to see into shadows, dark rooms, and at night. Because of the split vision, extended use of PVS-14s often gave the wearer spectacular headaches.

QRF (Kue'-ar-eff) **n. acronym** (Quick Reaction Force). A military force designated to be ready to respond to a necessary situation. Commanders use QRFs to handle crises as varied as IED strikes, casualty evacuation, or reinforcement of troops in a firefight. Generally, when on QRF duty, a unit sits in its vehicles, wearing its necessary gear and ready to move with a few minutes' notice.

Qunbula (Koom'-ba-la) **n. Arabic.** A bomb. The word used by Iraqi locals for IED.

Rear-D (Reer-dee') **n. slang** (Rear Detachment). The portion of a unit left behind when a unit deploys forward. The typical responsibilities of the rear-d include in-processing new soldiers, managing property and facilities, assisting with the medical treatment and funeral preparations for injured or killed soldiers, and dealing with family member issues and incidents of serious misconduct.

RPG (Arr'-pee-gee) **n. acronym** (Rocket-Propelled Grenade Launcher). A shoulder-fired, Soviet-era, rocket launcher which was widely used by pretty much every bad guy organization in the GWOT. The explosive projectiles it fired varied widely in effectiveness and accuracy, and they constituted a regular threat to personnel and vehicles throughout our time in Iraq.

RIP (Rip) **v. acronym** (Relief-in-place). the process of one unit replacing another during a mission.

Ripit (Rip'-it) **n.** A popular energy drink which was omni-present in Iraq and consumed in tremendous quantities by the soldiers there.

RTO (Ar'-tee-oh) **n. acronym** (radio telephone operator). A soldier whose primary duty was to handle radio communication for someone else, usually his leader. In a ground unit, the RTO was usually responsible for monitoring his higher headquarters' frequency to listen for guidance or coordinate with fire support, medical, or EOD assets, while his leader used the unit's internal frequency. In a command post, the RTO was a soldier responsible for monitoring radio systems while leaders were at meetings, asleep, or otherwise out of the area.

RUMINT (Roo'-mint) **n. abbreviation.** Short for rumor intelligence, which is a fictional type of intelligence that employs information collected from rumor and hearsay, usually through informal channels. It is famously accurate because it often avoids the hedging or softening to which published intelligence often succumbs. The term RUMINT is a play on the actual categories of intelligence, like SIGINT (Signal Intelligence) and HUMINT (Human Intelligence).

S-1, S-2, S-3, S-4, S-6. (Ess'-won, etc.) **n. abbreviation.** The S codes are the terms for the various functional offices that comprise an Army staff. The S-1 handles personnel and administration, the S-2 works intelligence and creative writing, the S-3 is the operations office, the S-4 does logistics, and the S-6 breaks all the communication equipment so nobody else in the organization can talk to one another.

VBIED (Vee'-bid) **n. acronym** (Vehicle borne improvised explosive device). A VBIED was a vehicle rigged with explosives. Sometimes they were detonated while stationary, by remote. Other times, they were driven to their targets and detonated by suicidal drivers. Because of the amount of explosive material a vehicle could carry, and the speed with which it could reach its target, VBIEDs were very dangerous.

XO (Ex'-oh) **n. acronym** (executive officer). The second in command of an organization who fills in for the commander when necessary and handles coordination of the unit's logistics. Also, generally the member of the command who does the stuff the commander doesn't enjoy.

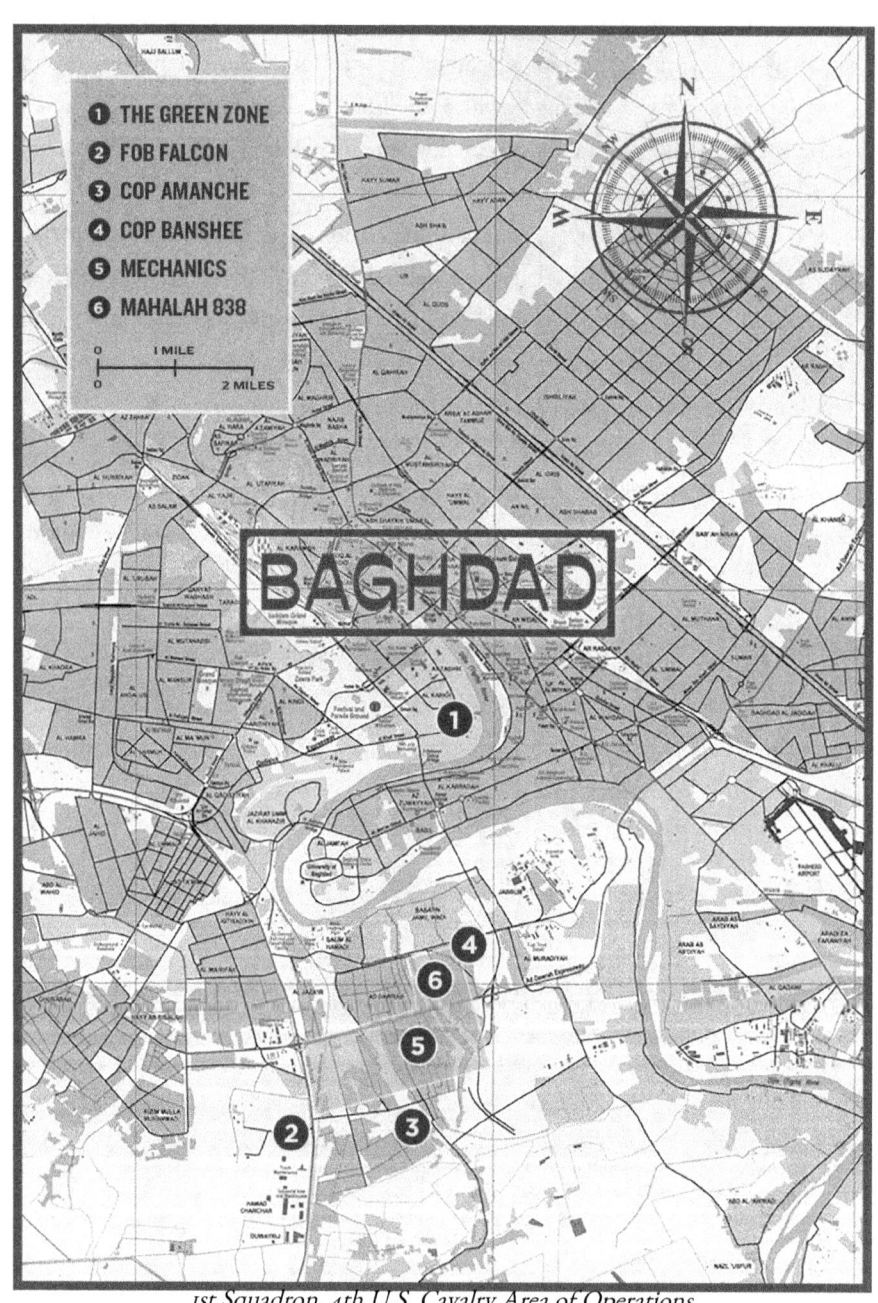

1. THE GREEN ZONE
2. FOB FALCON
3. COP AMANCHE
4. COP BANSHEE
5. MECHANICS
6. MAHALAH 838

1st Squadron, 4th U.S. Cavalry Area of Operations

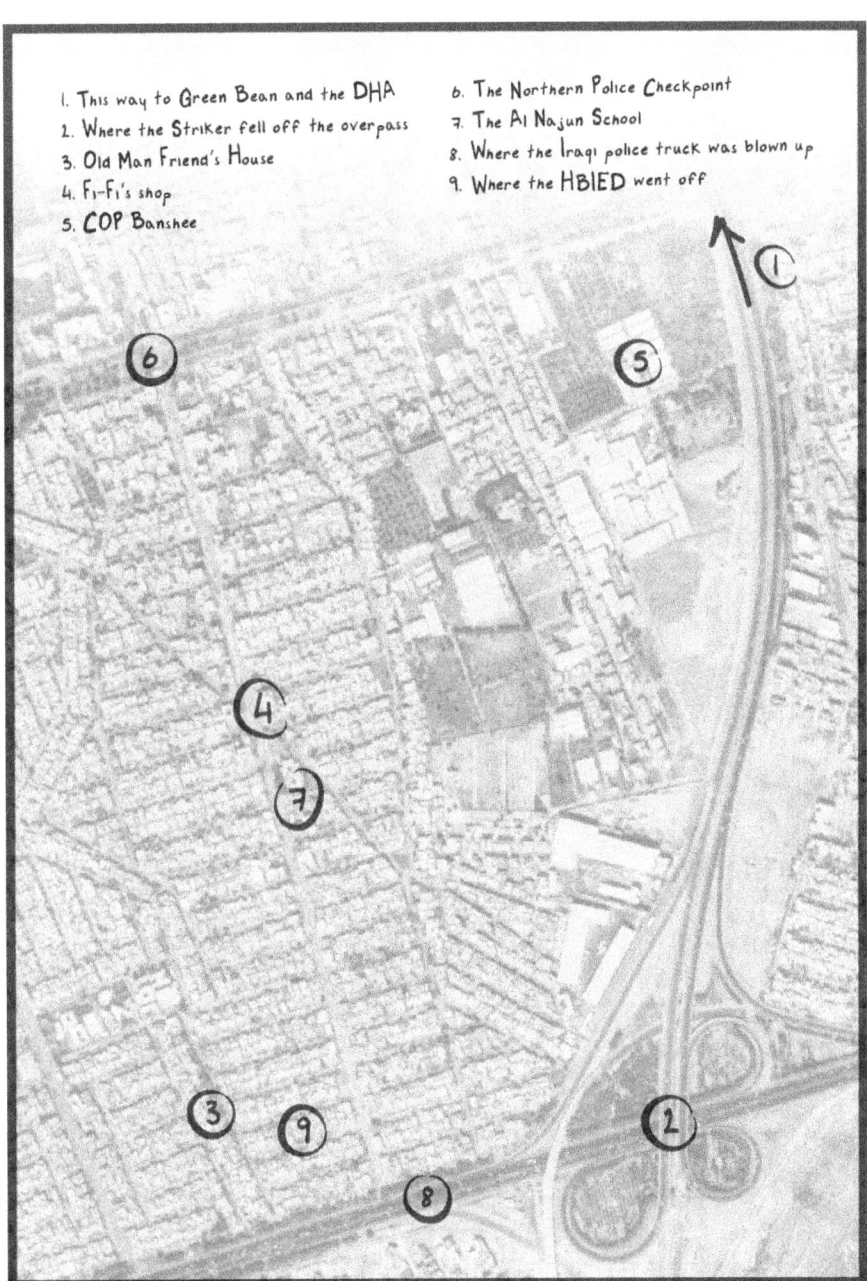

1. This way to Green Bean and the DHA
2. Where the Striker fell off the overpass
3. Old Man Friend's House
4. Fi-Fi's shop
5. COP Banshee
6. The Northern Police Checkpoint
7. The Al Najun School
8. Where the Iraqi police truck was blown up
9. Where the HBIED went off

Mahala 838 with highlighted points of interest

Forward

"War is hell, but that's not the half of it, because war is also mystery and terror and adventure and courage and discovery and holiness and pity and despair and longing and love. War is nasty; war is fun. War is thrilling; war is drudgery. War makes you a man; war makes you dead."

-Tim O'Brien, The Things They Carried

As a forty-three-year-old sitting at my computer and looking back on my time in Iraq with the 1st Squadron, 4th U.S. Cavalry, it's tempting to feel a bit gloomy. Thinking of the dead, wounded, and irreversibly changed friends and reading about the series of disasters that continuously unfold in the Middle East, it's easy to wonder if anything we did was worth the price we paid to do it. The funny thing is, though, when I actually read through my letters from that period or talk to the other guys who were there, I usually end up laughing about something.

That's because the events that took place when we stood up and went to war with the 1-4 CAV, whether terrible, hysterical, or ludicrous, were always so far outside any of our frames of reference, that it's impossible not to look back on them with wonder and amusement. And how could they not have been? We were all such a "locked, cocked, and ready to rock" bunch of young hooahs, that our time in Iraq couldn't have been otherwise. In the long-term strategic sense, I'm not sure what we accomplished, but whatever it was, for the fifteen months we were in Iraq, we accomplished the hell out of it.

One of the most interesting things about this story is that our unit's tale isn't even all that remarkable. As crazy, exciting, or tragic as some of our experiences were, they were the experiences of an entire generation of soldiers and family members, and almost everyone I know who served between 2006 and 2009 has similar stories about his deployment to Iraq. Most of the guys and gals I served with aren't writers, though, so the average American won't hear their tales unless he has a beer with one of them. By writing this book, I hope to give readers who didn't have the pleasure of deploying to Iraq a glimpse into our lives there and the experiences we had.

I have endeavored to make this book as true as possible. However, when I've tried to piece some of the events together over a beer with old buddies, I've discovered that whether because of my narrow perspective, a faulty brain housing unit, or maybe just general entropy, sometimes things don't line up quite right. In these situations, I've done my best to reconstruct these events as well as possible, making as few adjustments for style and readability as necessary. I handled the dialogue similarly. While all the conversations occurred, they didn't necessarily happen at the exact times or in the precise manner as they do in the text. To those who find this unsatisfactory, I apologize. Before they offer too harsh a judgment, I encourage them to spend a year standing in a sauna with sixty pounds of stereo equipment and ceramic cookware strapped to their bodies, have a hundred variations of the same conversation about generator maintenance with strangers in a foreign language, get repeatedly blown up by explosives, and then try to recount the experience in a readable manner.

This book is a memoir–history in its unpolished form–written by a person who lived through a distinct and, I think, interesting period of U.S. Army history. I leave it to professional historians to place this story in its greater strategic and historical context and to sift through the inconsistencies my limited perspective creates between this account and other people's. For me, it is enough to tell the story of how we built a unit from nothing, trained it on a shoe-string budget, deployed it to a war zone, and then spent fifteen months doing our best to accomplish a difficult mission and keep our guys alive. By telling this story, I hope to enrich the overall understanding of our small portion of The Surge, and to immortalize the contributions of the soldiers who gave up their youth, health, families, and lives doing what their country asked them to do. The 93 killed and over 700 wounded soldiers the Dragon Brigade suffered that year while trying to achieve stability in Baghdad deserve no less.[1]

---

1. Dragon, Task Force. 2008. Task Force Dragon: Operation Iraqi Freedom IV Yearbook. Baghdad: DoD.

Prologue

A Valley in Southeastern Afghanistan, October, 2003

"I have a picture,
Pinned to my wall
An image of you and of me
And we're laughing,
We're loving it all
But look at our life now,
All tattered and torn
We fuss and we fight – "

The metallic plink-plink-plink of machine gun rounds hitting the improvised armor of our Humvee door interrupted the sound of Tom Bailey's rendition of "Hold Me Now" that was blaring from the cheap set of speakers we had taped to the dashboard of our truck. The speakers were connected to a skip-proof Sony Discman balanced on the truck's radio, and as our gunner, E-Brake, jerked the turret right to return fire, his foot accidentally kicked the Discman under the radio mount.

Our radio blared to life with reports from the other trucks in our convoy.

"Killer-six, this is Killer-seven, contact, three o'clock."

Killer-six, our lieutenant, had the bubble guts today, so he was simultaneously vomiting out of his cracked door, keying the radio hand-mike that was attached to his helmet with tape and a plastic spoon to respond to Killer-seven, and fumbling for the Discman to turn it off.[1]

"Hold me now, whoa

---

1. The Army-issued radio hand mike of that era came with a small, useless clip to attach it to your equipment. Consequently, soldiers commonly taped the thick plastic spoon that came with their issued rations to the hand mike so they could hang it off their helmet strap, creating the 2003 version of hands-free technology.

Warm my heart
Stay with me
Let loving start"

"Turn that shit off!" he shouted, giving up his efforts to reach the CD player and wiping the vomit residue from his mouth with his sleeve.

From the back seat, I leaned around E-Brake's legs to grab at the Discman. My M203 vest, with its load of 40mm grenades, hindered my attempts, though, and I couldn't reach it. I shouted at the driver to get it instead.

"Sobon! Grab that shit!"

"So perhaps I should leave here
Yeah, yeah, and go far away
But you know that there's nowhere
That I'd rather be
Than with you here today."

Mirroring the truck in front of us, Sobon pulled our truck off to the side of the road so we could engage the tribesmen firing at us from the cliff overlooking our convoy. As he mashed the brake pedal, the truck lurched to a stop, and the Discman dislodged from underneath the radio mount where Sobon grabbed it and hit the pause button.

I slammed my shoulder into the perpetually stuck rear door of the truck. As it sprang open, I stumbled out of the truck, hanging onto the door to use its improvised armor as cover from the continuing fire. While the lieutenant shouted coordinating instructions to the other units in the convoy to flank the enemy element, I looked up at the cliffs. They were high and rocky, and the tribesmen were using large boulders for cover that my M4 wouldn't be able to penetrate, so I loaded a grenade into my M203 and fired it in a high arc toward the top of the cliff.

The grenade hit low, exploding violently against the cliff face, dislodging a rock, and startling a huge, black vulture which took to the sky with an irritated shake of its feathers. I reloaded the M203 and fired again, this time successfully angling the grenade over the boulders and hitting the top of the cliff.

The enemy fire stopped, and I scanned the cliffs for further activity. Our lieutenant had gotten out of the truck by now and started heaving bile into the dirt. "Man," I thought. "The Hajji crud got him good today."

"E-brake, you see anything?"

"Nope. No movement."

"You fresh on ammo?"

"I'm good so far, but can you hand me another can in case I need it?"

"Yeah, no problem."

I leaned over to unstrap the cans of ammo we had stacked at E-brake's feet nine days ago at *Orgun*. I got one of the heavy cans loose and passed it up, struggling a little to push the awkward, 49.2 pound, two-hundred round can through the turret hole so E-brake could grab it. Normally, .50 caliber cans only held a hundred rounds, but we'd rigged larger MK-19 cans from our other truck to hold an extra hundred rounds to cut down our reload times. I heard E-brake slam the new can onto the roof of the truck and felt a .50 cal link hit me on the shoulder as he swept the turret clear of the spent brass to keep the debris from catching in the turret's bearings and jamming its rotation.

"All good."

The lieutenant had recovered by now, and he looked back at me, smiling under his mirror shades.

"That was some good shit, Specialist Pace. We killed the fuck out of those douchebags." He looked down at the puddle of vomit at his feet. "I think I left a piece of my soul in the dirt, though. The tea and foot-bread that guy gave me yesterday are wearing my ass out."

The radio speaker confirmed the C Co. guys were clearing their way across the top of the cliff now. It would take them a while to finish the job, so we had time to kill.

"Yeah, it looks that way, sir. I never thought I'd have a firefight with the Thompson Twins playing in the background. That was crazy."

The lieutenant and I laughed at the weirdness of the situation. Of all the potential songs I could imagine for the soundtrack of a search and destroy mission in eastern Afghanistan, "Hold Me Now" wasn't on the list. Maybe a little Drowning Pool, or if you were in the One-Hundred-and-Worst Airborne, maybe even a little Peter Gabriel, but definitely not the Thompson Twins.[2]

As I continued to scan the surrounding cliffs from left to right and top to bottom, two old Afghan men walked by our truck, their sharp eyes peering out from under weather-beaten brows to meet ours as they passed. When our firefight had started, they'd both squatted down alongside the road to wait, and now that the firing had stopped,

---

2. One-Hundred-and-Worst being the pejorative 10th Mountain Division soldiers frequently used for their more glamorous cousins in the 101st Airborne Division.

they continued their slow walk from wherever to nowhere, shoulders stooped, and hands folded behind their backs as always. With no interpreter in the platoon, we had no way to communicate with them, but honestly, we weren't really interested in doing so. We were in this valley to find the sons of bitches that had attacked us on 9/11 and kill them, not rebuild this Third World shit hole. These people didn't want anything we had to offer, anyway. They just wanted to continue their pre-industrial age existence out here on the edge of the world without anyone messing with them.

Over the last few months, that last part had started eating at me. We were three months into our deployment, and it often seemed like all we were doing out here was screwing with a bunch of people who just wanted to be left alone. Every day, we drove the roads and humped the mountains of *Paktika* province, looking for Al-Qaeda terrorists, foreign fighters, or Taliban rebels, but our movements to contact were random, our harassment and interdiction fire purposeless, and our raids usually ineffective. We knew how to *fight* the enemy, in the sense that we were well-trained infantrymen who were good at killing people, but we had no idea how to *find* the enemy or to identify him within the sea of Afghans in which he swam.

I shrugged off these thoughts and returned my focus to the present situation. Today was the last day of our ten-day mission, and I looked forward to getting back to *Orgun*. The C Co guys finished their sweep of the cliff top, and Captain Lopez's voice came across the radio, directing the convoy to continue its movement. Our truck's tires crunched across the gravel as it rolled back onto the road to fall in behind the truck in front of us.

"Mind if we fire back up the tunes, Sir?"

The lieutenant finished a fresh round of vomiting and leaned back into the rolling truck. He closed his door, wiped his mouth again and took a swig of water.

"Absolutely, Pace."

I reached forward to set the Discman back onto the radio mount and hit 'play'. As the weak sunlight shone through the turret onto my gloved hands, Tom Bailey's voice resumed singing the '80s lyrics I was so familiar with. I stared out of the Humvee's window and wondered what the cooks had thrown together for chow tonight.

Part I

Assembling the Unit

Grey Zone Ethics LLC

I

Camp Darby, Georgia, November, 2005

My road to Iraq with the 1st Squadron, 4th U.S. Cavalry began and ended with Ranger School. They say everyone leaves Ranger School–the U.S. Army's premier leadership school–with either a tab or an excuse. In week two, after an aggressive encounter between my left shoulder and the stump jumper–a colorfully named station on the infamous obstacle course known as The Darby Queen–I left with the latter.

Ranger School is a wildly unpleasant experience. Much has been written about it elsewhere, so all I'll add here is that I'm grateful to the course for providing me with several of the most miserable days of my life, by which all other days seem pleasant in comparison. That might sound sarcastic, but it isn't. The Ranger School experience has provided me with a lifetime of context, and I'm (grudgingly) grateful to the course for that reason.

When I regained my senses, I was lying on the ground next to the stump jumper with an enormous, black-shirted instructor yelling at me to get up. I tried, failed, and then explained that my arm wasn't working. The instructor prodded my shoulder skeptically, noticed the collar bone was about four inches above where it should have been, and then changed his tone a bit.

"Stay there, Ranger." Then, turning his head, "MEDIC!"

Yada yada yada, the next day, I found myself driving from Fort Benning, Georgia, to Fort Riley, Kansas, in my 1997 Nissan Altima with a Class V acromioclavicular separation. I had to make the entire drive with the window down to accommodate the enormous cast the medics had put on my left arm. I can't remember if I was on painkillers or not, and I still do not know why the hospital released me to drive across the country. But since my car was in the Ranger School parking lot, flying wasn't really an option. Besides, the last thing I wanted to do was stick around Camp Rogers any longer than necessary, as the gentlemen in charge of out-of-cycle students had a passion for making the experience unpleasant. After a few days picking up trash and polishing brass at The School for Wayward Rangers, a fourteen-hour drive with just myself and Rand McNally was a welcome change of pace.[1] The drive itself was long but uneventful, and by that evening, I was pulling into our driveway, where my wife Alycia was awaiting my arrival.

Alycia and I had been married for four years, and our marriage was an Army cliché. Alycia and I had known each other growing up and on a whim, she had come to my basic training graduation in Columbus, Georgia, where we drunkenly agreed to get married over a boisterous dinner with my battle buddies at the LongHorn Steakhouse.[2] A few weeks later, we had a Justice of the Peace-moderated ceremony at a dive bar we'd both enjoyed in College Station, Texas, called Duddley's Draw.[3]

After our wedding, we moved to Fort Drum, New York, where I served as an enlisted man in the Army Infantry. One-and-a-half combat deployments and countless bizarre experiences later, I went to Officer Candidate School, graduated in May 2005, and received orders to serve as an Armor Officer at the 1st Squadron, 4th U.S. Cavalry at Fort Riley, Kansas. Grudgingly, Armor Branch let me attend Ranger School *en route,* I promptly failed, and I was now pulling up to my new home to recuperate a bit.[4]

Our house was a duplex on Lower Brick Row. It was built in 1889, when Fort Riley was still on the frontier. To say it was solidly built is an understatement. It was a square, brick structure with 18-inch-thick, poured concrete interior walls and decades of lead-based paint slathered on so thickly that the windows didn't open anymore. It had all the strange features of an old house that had been haphazardly remodeled on a limited budget. The house only had one bathroom, and it was on the top floor. Its single shower sat under low eaves, so anyone taller than 5'6" couldn't stand up straight, and it had a window at crotch-level that made showering an exciting prospect for the neighbors until the glass fogged up a bit. The basement probably used to be a root cellar or a frontier

---

1. For those readers of the GPS generation, the Rand McNally Atlas was the large, spiral-bound, Google maps of my youth.

2. As you might imagine, my booze-soaked proposal was legendarily bad. We still occasionally laugh about it today.

3. As of October 2023, the smoky old photo of that wedding is still hanging at Duddley's. Look for the lovely young lady standing next to the dope in a green suit with a bad haircut.

4. Armor branch is run by tankers, and tankers don't think much of Ranger School, but as a former infantryman, it was a school I very much wanted to attend.

bunker, but now it housed an enormous water heater, furnace, and washbasin, all of which looked like props from a horror movie. The house next door amplified the effect because it was condemned, sealed up, and featured prominently in the "Haunted Fort Riley" tour. Needless to say, our house hadn't aged terribly well, but standing in the living room, it wasn't hard to imagine pioneers crouching behind the windows firing at Sioux raiders.

Because I was injured, the Army sent me on convalescent leave. The surgeon was horrified that I hadn't been treated yet, and he scheduled a consultation and surgery within a few weeks. He did a great job (he even kept my tattoos straight while stitching everything back together), and I've never had trouble with the shoulder since. I spent Christmas recovering, and by January, I was in good enough shape to function as a soldier again.

Would I be allowed to function in the unit at all, though? I'd failed Ranger School, and in the infantry community I had been a part of for the last four years, that was a mortal sin, and no lieutenant arriving at my old unit without a Ranger tab would have been allowed to lead much of anything. They say everyone leaves Ranger School with either a tab or an excuse, and as I drove up Custer Hill on a cold Kansas morning to find the headquarters of the 1st Squadron, 4th U.S. Cavalry and meet my new commander, I wondered if my excuse would be good enough.[5]

---

5. Which should never be written or said as the ¼ CAV, unless you want a cavalryman to angrily point out that while there are quarter horses, the unit is always 100% cavalry–or "All CAV."

2

Fort Riley, Kansas, January, 2006

"There's no way this is the building I'm looking for."

I'd been in the Army for a few years, so I had some preconceptions about what the Raider Squadron headquarters would look like. In my experience, headquarters were usually buildings with well-manicured lawns, prominent unit heraldry, and fresh paint, and they invariably had sharply dressed soldiers manning desks inside well-polished foyers near the entrance somewhere. Their grounds were prowled by fearsome Sergeants Major, and their halls echoed with the mechanical roar of extra duty soldiers using floor buffers to shine the already-shiny floors.[1] After about twelve wrong turns and two stops to ask for directions, I found out the 1-4 CAV headquarters had none of those features, because it was in a converted chow hall.

The reason it was in a converted chow hall was because the 1-4 CAV was part of the 4th Brigade of the 1st Infantry Division, and the 4th "Dragon" Brigade was brand new. The wars in Iraq and Afghanistan were both burning hot, and the Army recognized that if it didn't want to continue deploying the same units and personnel repeatedly, it needed to grow. It accomplished this growth, while simultaneously moving toward a model which allowed smaller units to deploy independently, by creating the Brigade Combat Team (or BCT). The BCT pushed support assets that traditionally resided at the division level, such as intelligence, reconnaissance, and fire support assets, to the brigade level, and it expanded the brigade staff to coordinate these assets. The BCT also allowed the Army to expand and contract rapidly because its independent structure allowed it to be added to

---

1. Extra duty was a standard punishment for minor infractions among junior enlisted soldiers, like knocking your roommate unconscious with a keyboard because he touched your things or smuggling naked women out of your barracks window as the First Sergeant was pounding on your door before morning formation, to use a few examples from my time as an infantryman.

the existing division structure without overly taxing division resources. To accomplish the growth required to support the wars, between 2004 and 2005, most of the infantry divisions (including the 1st Infantry Division) received a 4th brigade, and the 1-4 CAV was one of the three maneuver battalions in the 4th Brigade, 1st Infantry Division.[2]

That all sounds very clean, but the reality was quite tumultuous. Army posts didn't always have the physical infrastructure to support this growth, so many of the 4th brigades found themselves shoehorned into rather unusual locations. The 4th Brigade, 10th Mountain Division, whose division sits in Fort Drum, New York, had to search so far afield for space that they ended up in the swamps of Fort Polk, Louisiana. We were luckier than that. Instead of ending up across the country in a mosquito-filled bayou, our BCT ended up in a set of unused buildings down the road from division headquarters on Custer Hill, Fort Riley, Kansas, which was where I was driving today.

After one last stop for directions, I pulled into a parking lot across the street from a building with a number placard that matched the one I was looking for. As I stared at the placard, mouth hanging open, I was sure something was wrong. The side of the building facing me was a loading dock designed for large trucks to unload equipment and supplies, complete with a rubber-bumpered concrete shelf and industrial railing. There were no signs, flags, or staff duty desks in sight. Assuming I was mistaken, but figuring it was worth checking anyway, I walked up the stairs and checked the double doors. They were open, so I walked in, wound my way through a few unoccupied rooms, and found an office where a few soldiers were sitting at computers.

"Hey, y'all. Is this the Quarter CAV?"

"There are quarter horses. We're All CAV," came the unanimous reply.

"Right, my bad. Is this the 1-4 CAV though?"

A sergeant looked up, saw that I was a butter bar, and gestured toward a side office announcing, "Hey, Sir, there's an LT here.[3] "

I walked toward the office the sergeant had gestured to, came to attention at the door, saluted, and said, "Lieutenant Pace reporting as ordered, Sir."

---

2. Government Accountability Office. 2005. *Force Structure: Preliminary Observations on Army Plans to Implement and Fund Modular Forces.* Washington, D.C.

3. The Army has slang terms for pretty much every rank. Specialists wear "sham shields" or are renamed as "full-bird privates," sergeants are called "buck sergeants," and First Sergeants are called "top." Second lieutenants, the lowest ranking officers in the service, have a particularly wide variety of monikers ranging from "butter bar," to "Louie," to "shit head."

For those readers who think this is a strange way for one adult to initiate a conversation with another adult, some explanation is probably required. During officer training, the Army inculcates two hundred years of tradition into its new recruits. During my flavor of officer training–Officer Candidate School, a short, twelve-week, crash course on how to be an officer–this indoctrination isn't reinforced by the years of context provided by the cadre in longer commissioning sources. Consequently, Officer Candidate School graduates often arrive at their units with little idea on how to actually function as officers in the Army.[4]

The Sir turned out to be a captain who turned toward me in his swivel chair, loosely returned the salute, and said, "Relax, LT." The captain was of medium height and build, with a prominent nose and a short, army-standard haircut. He had a combat patch, which meant he had been deployed, but no combat infantryman's badge, which meant he wasn't an infantry officer.[5] He looked me up and down for a minute with a set to his eyebrows which suggested something between disdain and dislike, then gestured toward the other computer in the room and said, "There's a personnel spreadsheet for new arrivals on that computer. Fill it out, and when you're done, go see the commander."

"Roger, Sir." I sat down, opened the spreadsheet in question, and started typing. Name. Social Security Number. Blood type. Pant and shirt sizes. Next of kin information. Allergies. Prior military schools. Deployment history. All the usual stuff. I'm the sort of person who fills awkward silences with awkward conversation, so as I sat back-to-back in this small office with this unknown person, I tried to make a bit of small talk.

"So, uh, Sir, do you know where I'm going to be assigned?"

There was a long pause.

"You're probably going to Alpha Troop. The commander will tell you when you talk to him."

There was another long pause.

"How does your family like the area, Sir?"

This time there was no response. As I clicked on the keyboard, I heard the captain

---

4. As a prior sergeant, I naturally met the officer indoctrination with a deal of skepticism, as I hadn't seen much of the formality the cadre described in my four years of service. "Maybe officers do it differently though," I thought. As it turns out, they mostly don't. But I didn't know that when I was checking into the 1-4 CAV.

5. In the Army, this sort of visual assessment ("butt-sniffing" in Army parlance) happens every time two new people meet. It's a side effect of wearing so much distinguishing information on our uniforms.

stand up and leave the office.

"Oh hell," I thought. "What did I say? Did I just ruin my reputation here on the first day? Not only am I a Ranger School failure, but now I've pissed off the first captain I bumped into."

I didn't know it then, but I'd asked almost exactly the wrong question. Like everyone else at the unit, the captain was a new arrival. As an armor captain, he'd been assigned to the 1-4 CAV to take command of Alpha Troop, but something went poorly during his interviews with the squadron leadership, so they assigned him to a staff job instead. He never ended up commanding a company during that rotation, and our relationship never really recovered from that initial awkward conversation.

The spreadsheet took another few minutes to complete. The beauty of most Army spreadsheets is that whatever information you don't know or don't feel like looking up can be readily invented without much impact. Spreadsheets are generally scrubbed for completeness, but not necessarily for accuracy. Occasionally this caused problems, like when you got boots that were three sizes too large, or when many years later a guy in my battalion answered the question, "If killed, what do you wish done with your remains?" with the reply, "Viking Funeral." He was killed in Afghanistan a few years later, but the command wasn't interested in loading him and his still-living wife into a boat and setting it on fire, so he got a normal funeral instead.[6]

After I completed the personnel spreadsheet, I stood up and walked through the headquarters building to find the commander's office. There was still almost no one in the headquarters, and almost nothing had been set up yet, so navigating the place was confusing. Eventually, though, I found the command suite and walked up to the commander's closed door. I knocked, and a voice inside said, "Come in." I opened the door and saw two men, Lieutenant Colonel Jim Crider and Major Craig Manville, sitting at a small table.

Jim Crider, the Commander of the 1-4 CAV, was a quiet, fortyish-year-old of medium height and slight build with a serious but not unfriendly expression. A thoughtful, intellectual man, the whole time I knew Crider, I never heard him raise his voice. Craig Manville was a more muscular guy whose friendly demeanor was punctuated by angry intensity when people screwed up, although I never had to be on the receiving end of his

---

6. During my time in the Special Forces, my own written funeral wishes were to have a friend named Al Gomez sing "Always Look on the Bright Side of Life" a cappella at my memorial. Fortunately, for both Al's and my wife's sakes, I didn't die before retirement.

wrath myself. As Crider's operations officer, Manville was often at his side, and I rarely saw one of them without the other.

I walked into the room, took a few steps forward, and saluted.

"Sir, Lieutenant Pace, reporting as ordered."

"Relax Lieutenant Pace. Have a seat."

I sat down at the table, unsure what to do with my hands. The Army teaches that when sitting in a formal situation, you sit with your back straight and your hands on your thighs. This is called "sitting at the position of attention." The presence of the table presented a confusing variable, though. My OCS sitting class hadn't mentioned tables. I put my hands on the table, palms down. I crossed my hands. I uncrossed my hands. I gave up trying to figure the situation out and left them where they were. Eventually, I realized that "don't look like an idiot" is an implied part of doing anything as an officer, but I initially struggled with that a bit, as most lieutenants do.

Crider spoke first. "Hi, Lieutenant Pace, how's the shoulder healing?"

"Good, Sir, the docs say they'll take the screw out at the end of the month, and I'll be back to full strength by March or so. Physical Therapy is going well."

The two men nodded. "Good. Well, you took the only honorable way out of Ranger School other than graduating. I hope you can make it back sometime."[7]

"Me too, Sir."

"Craig and I have been reviewing your file, and it looks like you were an enlisted infantryman with the 10th Mountain Division and that you already have a Combat Infantryman's Badge and a Bronze Star. Were you in Afghanistan?"

"Yes, Sir, Paktika province. 03-04."

"That's excellent. We'll be deploying in a year or so, and we'll need experienced guys like you to build the team."

Crider and Manville exchanged a brief look.

"You're going to Alpha Troop. As the most senior lieutenant, you'll be the executive officer. The new captains don't arrive for another month, though. Until then, you're in charge of the company. Don't do anything too crazy, just get down there and get the place running until the new commander arrives. The First Sergeant and platoon leadership teams are already there. Link up with them, and see what they need." Crider

---

7. There is a serious stigma in the infantry community against voluntarily withdrawing from Ranger School. In an infantry unit, having a "VW" scrawled on your end-of-course paperwork is the equivalent of having a scarlet A on your shirt in Puritan New England.

smiled. "Oh, and don't worry, I'll rotate you in to take a platoon eventually."

"Yes Sir, sounds great. Thank you."

"Don't thank me, Dan. Just go do a good job. You're married, right?"

"Yes Sir. My wife's name is Alycia. She's pregnant with our first kid."

"Congratulations, that's great news. Have your wife get on the Family Readiness Group contact list, and tell her my wife, Jill, will call her about the coffee group."

"Yes, Sir, I'll do that."

Crider and Manville stood up, indicating the meeting was over. I stood up as well and shook their hands.

"Welcome to the team, Dan. Get out there and do good work."

"Roger, Sir."

I left the office, walked out to the parking lot, and stood there for a minute. What had just happened? The news that I was going to be an XO, and *briefly a company commander*, was surprising and a little disappointing. Second lieutenants take platoons. Eventually, senior lieutenants become executive officers, but only after a year or two of work. As a prior sergeant, I didn't know what an executive officer was supposed to do, and I had no idea how to command a company. I'd been hoping to take a platoon so I could figure out life as an officer in a job I understood, but now it looked like I'd have to learn a lot more than I thought.

"So, I'm a company commander?" I thought. "What the hell would my old platoon think of that?" I then realized I didn't even know where Alpha Troop was. I looked around blankly at the unlabeled concrete buildings which surrounded the parking lot, hoping to find a clue of some sort that would point me in the right direction. Unfortunately, I couldn't find any. I noticed a group of soldiers standing next to a long row of buildings thirty yards away, though, so I swallowed my pride and walked over to ask them for help.

"Yep," I thought, "one more lost LT in the world."

As I walked up, the three of them turned to face me, put down their cigarettes, and saluted. One of them was a staff sergeant, and the other two were buck sergeants.

I returned the salute, and as I dropped my arm, I looked more closely at the short, intense looking staff sergeant and realized I knew him.

"Jeff? Is that you?"

He looked at me blankly for a minute, then his face broke out into a grin as he recognized me. "Holy shit, Sir. When did they make you a lieutenant?"

"I guess even a blind squirrel finds a nut sometimes. I haven't seen you since basic training back in Georgia. What the hell are you doing here?"

"I got assigned to the 1st ID after basic, and I've been doing the mechanized infantry thing since 2001. I went on the last trip with the brigade up the street, but I got the shit shot out of me by a tank in Iraq, so they patched me up and reassigned me here to teach these dipshits how to fight a war."

He rolled up his pant leg to reveal a long row of thick, puffy stars of scar tissue that ran from his calf to his mid-thigh. He continued his description.

"I was in a sniper position overwatching a route, and one of the tanker dickheads got trigger happy and lit me up. The only thing I was pissed about was not getting to kill any more fucking terrorists before I got home." Jeff punctuated his story with a wolfish grin. He had the look of a guy who meant what he said, and scars aside, I didn't think he was just blowing smoke.

"Damn, Jeff. That's quite a story. What do you think of the unit, so far?"

Jeff gestured to the two other NCOs standing with him. "You're looking at it, Sir. There ain't no unit to speak of. Shit's fucked up."

He paused and blinked, his mind appearing to wander off a bit before he continued.

"I can't wait to get back to killing motherfuckers, but I'm not sure how to do it with this shit here."

I didn't know what Jeff meant exactly, but he seemed uncomfortable talking about it, so I didn't want to dig into the problem.

Nodding, I changed the subject. "I'm not even in the infantry anymore. They branched me into the cavalry, so I'm going to work as an XO in Alpha Troop."

Jeff laughed. "Ha! Pussy! You're going to get all soft working with them tankers."

I laughed. "Maybe so, Jeff. Maybe so. You know where Alpha Troop is, by the way?"

He pointed at the long, low building behind us. "Yeah, A, B, and C are in that building. I'm in C with the rest of the infantry badasses. A's at the other end of the building. Just look for the fat people."

I laughed again. Old prejudices die hard. "Thanks, Jeff. I gotta get over there and see what's up. It was good seeing you."

Jeff stared at me for a few seconds, puffing on his cigarette. "You too, *Sir*. I still can't believe they made you a fucking officer."

"Me neither, man."

I walked past Jeff and his fellow infantrymen and crossed the parking lot to find Alpha

Troop, thinking about what Jeff had said. Was our unit screwed up? What did he mean when he said, "You're looking at it?" What the hell kind of organization was this?

"Ah well, I guess I'll find out."

The mostly empty building smelled of metal, mold, industrial lubricants, and pine oil. Alpha Troop–or Apache Troop as it was eventually renamed–was based in a long, one-story building, across the street from the squadron headquarters.[8] It shared the building with the two other 1-4 CAV maneuver elements, Bandit Troop–another cavalry element–and Comanche Troop, the squadron's infantry company. Each troop's portion of the building had its own entrance, a small area with offices for the troop leadership, and a row of offices for each of the three platoons. Behind these were a series of cages for equipment storage, an arms room, and a few small classrooms. Each room had a few sticks of furniture but was otherwise empty.

I walked in through the unlocked back door, didn't see anyone, and proceeded through the equipment storage room to the company headquarters area, where First Sergeant Richard "Dick" Strong sat at a long conference table talking to six other guys.

In the Army, every company-sized element has a First Sergeant. The First Sergeant is a central, fatherly figure, who serves as both the senior advisor to the commander and as a mentor to everyone in the organization. As with real fathers, some First Sergeants are domineering totalitarians, some are disinterested drunks, but most are great people who improve both their organizations and the lives of the people with whom they associate. First Sergeant Strong was one of the latter. An older guy of medium height, Strong had an ever-present dip of tobacco in his upper lip, a dry, Socratic method of questioning people and leading them to the right conclusions, and a generally easygoing demeanor

---

8. I voted for Aardvark company myself, but the Army heraldry folks aren't too keen on naming units after weird-looking animals.

which could turn on a dime when a situation called for someone to be "The Ass-Man."[9] The six people he was addressing were the leaders of Apache Troop's three platoons.

1st Platoon was led by Lieutenant Rob Humphrey, a short, calm, competent officer whom I never saw frazzled in any situation, and Sergeant First Class Troy Murray, a huge, bald, Southern guy with a prominent gold tooth, a big personality, and a hot temper. They couldn't have looked or acted more different, but they were a solid team who played off one another's differences well and used their divergent personalities to solve problems effectively.

Lieutenant Oroch "Koky" Sisoura and Sergeant First Class Gannon Edgy led 2nd Platoon. To this day, Koky remains one of the best-natured people I've ever met, and he and Edgy, an exceptionally intelligent Silver Star recipient from the invasion of Iraq in 2003, ran a solid team.

3rd Platoon was run by Lieutenant Alex Torres, a tall, confident guy from South Carolina with plenty of bravado, and Sergeant First Class Devon Winnegan, an older NCO with a sharp eye for detail and an understated personality. Torres was loud and personable whereas Winnegan was quiet and thorough, but between them they ran the most disciplined platoon in the company.

The assembly turned to look at me when I walked up. Strong immediately asked, "You the new XO?"

"Yep, that's me, First Sergeant."

"Great. I'm tired of signing shit. C'mon over and sit down. We're in the middle of the weekly meeting."

I walked to the table and sat down next to Strong as he continued the meeting. The topic of conversation today was the company's personnel. On the one hand, it was an easy conversation, because aside from the people seated around the table, we didn't have any–apparently that's what Jeff had meant outside when he said, "You're looking at it." On the other hand, integrating all the new people when they finally did arrive was going to be tricky, and that's what the conversation today was about.

---

9. "The Ass-Man" is a term coined by another First Sergeant I once worked for named Uncle Mike. He had a different but equally effective approach to leadership. Uncle Mike's philosophy was that there was no soldier problem that couldn't be solved by healthy dose of extra duty, and "You're working this weekend!" was a frequently heard comment at Monday morning formations. Reasons for weekend duty were varied but included smelling like booze, having a shaggy haircut, showing up in the wrong uniform, or simply having a face that looked like some extra work might improve it somewhat.

From the Army's point of view, one of the challenges involved in standing up a new battalion is figuring out how to bring new people into the organization. The bulk of an 80-man element like Apache Troop consists of junior enlisted soldiers–privates. The Army can produce these pretty quickly and in large quantities, so filling a company with privates is relatively straightforward. Privates can't be left unattended, though, or bad things happen, so before sending sixtyish privates to Apache Troop, the Army needed to ensure enough leaders were present to manage the new arrivals.[10] The eight guys sitting around the table, minus the captain who hadn't arrived yet, represented the senior leaders of the company, but Apache Troop still didn't have any of the middle management it needed to function. Without the sergeants and staff sergeants necessary to help the platoon leader and platoon sergeant manage the eighteen-man cavalry platoon, effective combat operations would be impossible. Since there weren't enough cavalry sergeants in existence, the Army's plan to address this problem was to retrain junior NCOs from other jobs which were in low demand during the GWOT, such as air defense artillery and combat engineering, add them to the mix with the senior leaders and privates, and trust that a year of training together would work out all the kinks.

Whatever confidence the Army may have had that this process would go well was not shared by the people in the Apache Troop conference room that morning.

In his thick Mississippi accent, Murray articulated what everyone at the table was thinking. "First Sergeant, what the hell are we going to do with a bunch of reclassed duck hunters? They don't know shit about being in the Cav, and we don't have any equipment to teach them with."

Murray was right. In addition to lacking personnel, Apache Troop had no weapons, vehicles, or equipment.

Strong had a ready reply. "Don't worry, Murray, the new XO is going to get us all our gear in a hurry. Right, Sir?"

Strong and Murray both looked at me. Their faces did not suggest confidence in my equipment-getting abilities. At least I knew what an XO was supposed to do now, though.

---

10. Bad things can happen indeed. In 2001, when I was a specialist, a group of buddies and I were sitting unattended on an airstrip in Uzbekistan, and we somehow managed to talk ourselves into suspecting we had symptoms of nerve agent poisoning. We were reading the instructions on our atropine autoinjectors when a sergeant walked up, told us we were idiots, and put us to work. He saved us from whatever the negative health consequences of atropine injection are, and he taught me a valuable lesson on just how dumb groupthink can get.

When in doubt, I thought, act confident. "Absolutely, First Sergeant."

Apparently placated by my response, they turned back to their conversation, and Strong continued. "As far as the soldiers are concerned, the Sergeant Major says we'll have our troopers in a few weeks. Knox just finished a class, and there is a whole pile of them headed this way. On the NCO side, I don't seem to have any spare Cav scout E-5's or E-6's stuck up my ass, so we'll have to make do with what we've got."

Murray nodded in reply, clearly unsatisfied, but the First Sergeant's point was valid. What else could we do but make do with what we had? The rest of the meeting was an unremarkable review of normal troop business, and when that was complete, First Sergeant adjourned the meeting. The platoon leadership teams filed by and shook my hand, then went back to their offices. When they had left, Strong turned to me, gave me the Army once-over, and started talking.

"How's the family getting settled in, Sir?"

"Good, First Sergeant, we're living down the hill on Old Post. My wife is getting the house together."

"She working?"

"No. She's due with our first kid in May, and she wants to stay home with the baby."

"Ah, good. Congratulations. So, it looks like you've been overseas, yes?"

"Yep. I was in Afghanistan with the infantry. I was a sergeant before commissioning."

"Good. Then you won't be as stupid as the other lieutenants, although I can't say I think much of your choice to go to the dark side!" He smiled at this remark and continued. "I also can't see why the hell you'd want to go to Ranger School. Shoulder healing up?"[11]

"Yeah, it's getting better."

"Good. Get better. The new commander will be here soon. His name is Captain Cook, and I've worked with him before. He's a good guy, and we'll be in good hands. Until he gets here, get our property straight, get us equipment to train with, and sign all that shit over there on the desk." He pointed at a foot-high stack of paperwork. "Your office is over there. When I need more shit signed, I'll put it in your 'in' box."

I turned to look at a pile of staffing folders on the nearby desk and nodded.

"Okay, First Sergeant. I'll knock that out."

---

11. The cavalry has an aversion to anything that smells too much like the infantry. This includes both Combat Infantryman's Badges and Ranger School. "Going to the dark side" is slang for an NCO switching to being a commissioned officer.

He nodded, shook my hand, and said, "Good to meet you, Sir. Formation is at 0630 every morning in PTs," before walking back to his office. I turned to the pile of paperwork, picked it up, and walked over to my office to sort through it. It was a pile of personnel actions. Each of the actions was a five-or-six-page block of text marked in the appropriate area with a yellow "sign here" sticky note. Being new to the officer world, I hadn't yet realized that reading all the words (rather than just the subject line) on an Army personnel action is a waste of time, so I spent a half hour trying to digest the bureau-speak and gobbledygook before signing the necessary blocks. I walked the documents over to the First Sergeant's office, put them in the "in" box next to his door, and left to find the other lieutenants.

When I found them, they were together in an office immersed, in a conversation about how their weekend in Manhattan had gone, and they all waved me in enthusiastically.[12] As I entered, Torres was in the middle of loudly recounting one of the weekend highlights. "So then Meech walks in the house, staggering drunk, and starts waving his arms around. Somebody hands him a drink, he pounds it, then falls over backward onto the coffee table. The trouble is, there's a set of wine glasses on the table, and he lands right on them. Of course, they shatter, and Meech spends the entire night in the E.R. getting his taint stitched back together. Look, he sent me a picture of that shit!"[13] We all leaned over. Sure enough, the tiny display on Torres's flip phone depicted a photo of a hairy, stitch-filled ass.

It took us several minutes to recover, but once we finally stopped laughing, we decided to head to lunch. Humphrey had a new silver Mustang and offered to drive. We took him up on the offer and made the half hour drive east to have some bang-up Kansas BBQ. After lunch, I headed back to the office, hung up a whiteboard and a set of markers that were sitting on my desk, and thought about the way ahead.

The command portion of my job seemed like a paper drill; I was just a placeholder, and there wasn't much to command right now, anyway. My real responsibilities as XO were going to be finding equipment, weapons, and money for the troop. We didn't know exactly how much time we had to work with, but RUMINT suggested we'd

---

12. The Little Apple, not the Big one.

13. Meech was the nickname of one of our battalion fire support lieutenants. He lived a wild, single man's life in Manhattan, and his weekend exploits were routine topics of conversation on Monday mornings.

be deploying in a year or so.[14] Planning backwards from then meant we'd have a month-long Squadron training event at one of the national training centers before we deployed, and that meant we'd have to conduct troop level, platoon level, and individual training before that happened. Before we could even start training, though, we needed equipment and people, and I had no idea how long getting them would take. As I stared at the whiteboard, sketching out the way I thought things would play out, it seemed like there was no way we would have enough time to do everything necessary to get the troop ready for war.

---

14. RUMINT, or rumor-based intelligence, is the speediest and most effective form of intelligence collection in the military. Douglas Adams asserted that nothing travels faster than bad news, but I believe RUMINT could give it a run for its money.

3

Fort Riley, Kansas, April, 2006

Custer Hill is one of the few places I've been where the weather is windy all the time. In the summer, it's hot and windy; in the winter, it's cold and windy; and in the spring and fall, it's wet and windy. Sometimes the wind even gets entirely out of line and spawns colossal tornadoes. Even when it's not tornadic, though, the wind on Custer Hill has been an omnipresent and often unpleasant feature of life at the 1st Infantry Division for as long as it's been at Fort Riley.

Today was no different. It was 0629 on Monday morning, the sun was just thinking about rising, an icy wind blew steadily out of the north, and Apache Troop was standing in a rectangular formation waiting for the cannon to fire. The cannon fired every day at 0630. Its discharge signaled the start of the duty day, and it was always immediately followed by the playing of Reveille, the Star-Spangled Banner, and the 1st Infantry Division song.

Dressed in their grey and black PT gear, reflective belts gleaming, the men of Apache Troop saluted smartly through the first song and sang the second with a degree less enthusiasm, surging only to maximum volume to shout, "pride of the *cavalry*!" over the bit at the end where the song's actual lyrics were, "pride of the infantry." Afterward, First Sergeant Strong announced last night's DUIs, this morning's urinalysis victims, and this afternoon's meeting schedule. He then turned the formation over to the platoon leaders to lead PT. As the XO, I could accompany whichever platoon I wanted, so I joined 3rd Platoon for hand-to-hand combat training. I fought a couple of guys successfully, but then lost to Major Manville, who joined us that day and promptly bent me up like a pretzel.

After PT, I drove home for a shower and breakfast and thought about the day's tasks. Organizationally, Apache Troop had started coming together over the last few months.

Captain Nick Cook, a fiery armor officer who had served with the 1st Infantry Division in Iraq a few years before, was in charge now, and he had teamed up with Strong to whip the troop into shape. The platoon leadership was gelling as well; the platoon sergeants were getting a handle on how to work with Strong, and the lieutenants (myself included) were learning how to avoid pissing off the new commander too much with our butter bar antics.

Our much-needed sergeants and staff sergeants had finally started arriving, too. All three platoons received a senior cavalry staff sergeant, each of whom had just come off a combat rotation, and a collection of other NCOs from all over the place. As predicted, most of them were re-trained sergeants from the military police, engineers, and air defense artillery. Some had combat experience, and some didn't. Most were hard working, effective leaders, and a few weren't. All of them were new to both their rank and job, and much work would be needed to forge them into a fighting unit.

Finally, one April morning, our soldiers arrived *en masse*, and they were exactly what privates have been throughout American history. We had rich kids—one of them regularly vacationed at The Moonlite Bunny Ranch in Nevada—and we had kids so poor they'd shown up at the recruiting station with nothing but the clothes on their backs. We had all-star high school athletes, and we had a guy who'd needed to lose so much weight to join the Army that he'd needed surgery to remove the excess skin afterward. We had ex-convicts and fresh-faced guys so naïve they'd never left their hometowns. We didn't know it, but we even had a guy who was such an alcoholic that he would eventually drink his way straight through a mandatory Anabuse prescription and be found by the military police on the golf course in a pool of his own vomit and urine with a .27 blood alcohol level. Our cavalry troopers were a motley crew. They came from all over the country and from every walk of life, and their only unifying feature was that they were standing together on this chilly Kansas morning with the certainty in their minds that 1SG Strong would kill them if they weren't.

Our troop had finally started to receive equipment, too. One challenge of building a unit during The Surge was the way the Army started handling equipment distribution. While cutting-edge individual gear, like weapons, optics, night vision, and radios, was promptly issued to our unit, the Army now provided unit-level gear, like vehicles, to units as they were rotating into combat. This allowed the Army to ensure that units arriving in combat had the most up-to-date vehicles available without having to buy them for and transport them from every base in the U.S.A. For established units, this wasn't an issue, as they already had old equipment to train with. For a new unit, though, this presented a major challenge: how do you train an organization to conduct combat operations in vehicles when you don't have any vehicles to train with?

That was the problem I was chewing on as I drove home that morning. When I got to the house, Alycia was in a bathrobe, stirring a pan of scrambled eggs on the stove for breakfast. She was eight months pregnant now and wildly uncomfortable doing pretty much anything. The pregnancy was going well, but she was very ready to not have a human being inside of her anymore. We talked over breakfast about the business of the day, and afterward I headed to work. Today was the Squadron XO meeting, and I needed to get my ducks in a row before I went.

Our Squadron XO was a guy named Major Rick Moon. He was an old tanker with a head of white hair, a bottle of scotch in his desk, a love of the cavalry, and a ruthlessly detail-oriented personality. I once got to spend an evening redoing a walkthrough of our motor pool because I forgot to draw a circle around an "x" on a maintenance form. Moon had turned to me and said, "Dan, I could just let you circle this now, but then I wouldn't be teaching you anything." He had paused for a moment, thinking, then continued, "I've got a wife, a dog, and kids. I don't need any friends. Go redo the walkthrough." His personality was exceptionally well-suited to getting the unit ready to deploy, but he wasn't always a fun guy to work for.[1]

When I arrived at the office, I studied my whiteboard to see which training events I'd be talking about during the meeting. Rifle range–no issues. Radio training–no issues. 3rd Platoon heavy weapon range–that would provoke questions. Suddenly, a thunderous voice shouted, "XO!" Captain Cook was apparently in the office. His office was next door to mine, and shouting "XO" was generally the way he got my attention.

"Yes, Sir!?" I shouted back.

"2nd Platoon is coming back from the range at 1600! Make sure Long's got the arms room open and that he inventories the weapons when they get back!"

"Roger, Sir!"

We had whole shouted conversations like this sometimes. I gave the whiteboard a last look and started walking up to squadron headquarters for the XO meeting. On the way, I dropped by the arms room and saw Braden Long, our armorer, taking one of the

---

1. His lessons on the importance of paying attention to detail have never stuck with me as well as they should have, much to Alycia's chagrin.

M249 SAWs apart. At 18, Long was one of the youngest guys in the company. He was a quiet, red-haired kid with glasses and an easy smile, and he'd been unlucky enough to be tapped to run the arms room almost immediately after arriving at the troop.[2]

"Hey, Long, how's the arms room running today?"

"Good, Sir."

"You tracking that 2nd Platoon is coming in this afternoon?"

"Yeah, I am, Sir. They talked to me. I'll have the door open."

"Great, you're all over it. Don't let them bullshit you, man. No missing slings and no dirty weapons, or you and I will be cleaning them ourselves before the boss walks through next week."

"Gotcha, Sir."

"Awesome. Later, Long."

I left Long to his repair work and walked across the parking lot to Squadron. In the last few months, the headquarters had come a long way, and most of the offices which had been empty before Christmas were now full of bustling staff guys. On the way to the conference room, I swung through the S-4 shop to say hi to Lieutenant Mike Castillo, the squadron logistics officer, and to see if he had any business to work through before the meeting started.[3]

"Hey, Mike, how're things?"

"How do you think they are, Dan? I spend my days with the Squadron XO's hand up my ass like a sock puppet."

We both laughed. Mike was also old for a lieutenant, and being a former NCO, shared my barracks sense of humor. He was an armor officer like me, but he'd shown up at the unit a few months later than the other lieutenants, so he was put into a staff billet. If there's such a thing as positive karma, Mike's time in the 1-4 CAV S-4 shop certainly earned him a bit.

"Do I owe you anything?"

---

2. Usually, the armorer is a senior soldier who has either finished his time in the Army and is on his way out or is recovering from an injury and can't train with the platoons anymore. It's a thankless job. You work long hours, you're responsible for millions of dollars of equipment, and everyone thinks you're a jerk because you're the guy who makes them clean their weapons before they get off work. The 1-4 CAV had no senior soldiers, so we just pulled the most organized looking guy we could find and hoped for the best.

3. One lesson Mike and I learned early from MAJ Moon was that if you needed to talk through the details on something you disagreed about, you had better do it before his XO meeting, or he would use your ill-preparedness as a reason to dig into both of your asses.

Mike ran his fingers through what was left of his hair and looked at his whiteboard before replying.

"Do you have the fuel request submitted for the heavy weapons range yet?"

"Yeah, I pushed it through Lieutenant Mattison."

"Okay. You have med, range control, a water buffalo, and everything else?"[4]

"I'm tracking there are no more buffaloes left, so we'll need water cans dropped."

"Alright, I can work that." Mike scribbled some notes in his notebook and looked back up. "It's about that time, Dan. Ready for the meeting?"

"Yeah, let's do this."

Mike and I walked to the squadron conference room, where the B Troop XO, Brent Warren, Headquarters Company (HHC) XO, Matt Babiarz, and Forward Support Company (FSC) representative, Mattison, were already sitting around the big, square table. Matt was a West Point graduate, and he was animatedly telling the group a story about his experiences in the freshman gymnastics course there.[5]

"So, at West Point, everyone has to take freshman gymnastics. Even those of us who aren't naturally inclined that way." Matt paused to indicate his stocky, 230-pound frame. "You know what this shit looks like in a unitard? Can you imagine me hitting the pommel horse? It wasn't a pretty sight, I assure you."

We were all still laughing when Major Moon walked in. The laughter died, and we all stood at attention.

"Carry on, gentlemen. Where's Comanche Company?"

We all replied with blank looks and unhelpful shrugs. Comanche was the squadron's infantry company, and from the beginning they had been the black sheep of the squadron. Their commander was an exceptionally effective and charismatic guy named Captain Laybourne. He had the company training hard, but he and his NCOs (like my old friend Jeff) ran a very unorthodox organization, and that, combined with his XO's frequent tardiness, ensured MAJ Moon was regularly irritated with them.

Moon nodded to himself and said, "Let's get this started." Mike Castillo walked up to

---

4. Once upon a time, before anyone sold bottled water or energy drinks, Army guys used to drink nothing but coffee at work, and the Army used to provide water for training events in trailer-sized metal water drums called water buffaloes. Every training event required either one of these buffaloes, its smaller cousin, the hanging lister bag, or a pile of five-gallon cans full of water.

5. Brent Warren was the biggest guy in the battalion. Not coincidentally, he was chosen to judge the cake contest at the January battalion social. Incendentally, Alycia's cake won.

click the slides, and we ran through them together. The meeting went relatively smoothly. I would have gotten hemmed up on the water issue, but thanks to the pre-meeting,
Mike was tracking, so Moon was appeased. Brent Warren was lectured on the importance
of paying attention to detail because of a slide error, and when Lieutenant Fedish, the
Comanche Company XO, walked in halfway through the meeting, he got an earful on
timeliness and professionalism. Finally, we got to the last bit of the meeting, Lieutenant
Mattis's maintenance slides. These were MAJ Moon's favorite to chew on. He had come
from a tanker unit, where maintenance was king, and parts always arrived promptly.
The 1-4 CAV was different, and parts for the few aged vehicles we had on hand were
almost always late. Mattis, and to a lesser extent the rest of us, regularly paid for Moon's
dissatisfaction with the situation with our time and patience.

As the meeting wrapped up, we stood up and saluted, then went our separate ways.
The misunderstandings and disagreements that came up during our XO meetings were
common. It would be easy to blame them on the Squadron XO's rigid personality, just
like it would be easy to blame them on our own inexperience, but the problems were
baked into the 1-4 CAV's structure. Our squadron was a light reconnaissance unit, attached to an infantry brigade, and it was manned by a combination of armor and infantry
officers. The cultural divides between heavy tankers and light cavalrymen, between the
old army and the new one, and between the infantry and armor communities were
deep, and moreover, there wasn't consensus among our leaders about how the squadron
should accomplish its assigned mission or what reconnaissance operations should look
like during the GWOT.

Before arriving at the new unit, the leaders of the squadron had vastly different
experiences and correspondingly different opinions about how the squadron should
train for war. Some of us had served in the rural mountains of Afghanistan, while others
had served in the thunder runs that overthrew Saddam's army in Iraq in 2003 or the two
years of occupation that had followed. Some of us came from the light infantry, where
reconnaissance work meant camouflaged foot movements and long days staring through
rifle scopes from hidden positions. Others came from the heavy armor community,
where reconnaissance meant moving rapidly to stay ahead of the tank battalions and
fighting for information. These experiences were completely different, and they had
burned correspondingly different lessons into our respective brains. Layered on top
of this all was the tendency of junior leaders like me to bump heads with the old
army crew. We brought home lessons from multiple combat deployments on how to

train and what equipment to use that sometimes contradicted the traditional ways of doing business, and many senior leaders were unenthusiastic about implementing our recommendations.

All of this, combined with the lack of institutional knowledge in our new squadron, made friction inevitable, and this friction sometimes manifested itself as frustration or in unmet expectations, like it did in our weekly XO meeting.

I thought about all of this as I walked back across the street, where I found Long managing the weapon turn-in for 2nd Platoon. Sure enough, a guy had lost a small pin for his rifle while cleaning it and was trying to talk Long into letting him turn the weapon in without it. I interrupted the conversation.

"What's up, Long?"

"2nd Platoon is ready to turn in, but one of their guys lost a firing pin retaining clip while cleaning his weapon. I've got a box of them in here; we can replace it with no issue."[6]

"Nope." I turned to the young NCO and said, "Tell your guys to find the pin. I don't care if it takes until midnight. We don't lose weapon parts."

The sergeant nodded and walked back to his guys to break the news. They were clearly pissed, but they started looking for the pin. I walked back to my office and sat down to wait for them to find it. I had once spent most of a weekend at Fort Drum with my entire company searching the snow-covered land navigation course for a lost piece of an M203 grenade launcher. It had fallen off some guy's weapon while he was checking his map, and Uncle Mike had made it abundantly clear that we would not go home until we found the piece. Sure enough, after about thirty hours of trudging through the knee-deep snow, we'd found it, and the unit didn't lose a weapon part again the entire time I was there.

A few hours later, 2nd Platoon found the pin, and we all headed home. I laughed to myself on the drive home and thought that maybe some things about the old army should be kept around after all.

---

6. A firing pin retaining clip is a particularly small and inexpensive piece of a soldier's rifle. It is perfectly shaped to fall into crevices and holes of almost any size, and it's (of course) absolutely critical to making the weapon fire.

4

Interstate 70, Kansas, July, 2006

The ride from Fort Riley to Abilene, Kansas, the hometown of President Dwight Eisenhower, took about an hour. It was a pretty drive that July day, and as we headed west on I-70, I watched a huge storm from the Rocky Mountains roll across the plains, rippling the grass in waves like a wheat-colored ocean. The weather was hot, but fortunately for the officers of the 1-4 CAV, the bus we'd taken from the motor pool for the trip was equipped with cutting-edge army air conditioning units.[1] Every officer in the squadron was on the bus, dressed in the ubiquitous khaki-pants-polo-shirt combo that officers are required to own and wear on these occasions.[2] Some, like the Apache Troop guys and the other XOs, I knew well, but others, like the Comanche Company platoon leaders and the squadron artillery officers, I had never interacted with much. We were spread across the bus in informal clumps and enjoying time away from the squadron as we sweated our way across Kansas to better ourselves professionally by learning about the life and career of one of the nation's most famous military leaders.

The relative quiet on the wind-blown bus was a welcome diversion. Life at home over the last few months had been hectic. In May, Alycia had given birth to our son, Gareth. Like all first-time parents, the actuality of the birthing process had overwhelmed us, although it had gone smoothly in the end. Gareth was a fussy baby, though, and we never found a remedy, so his waking hours were a constant struggle to keep him happy, which had made learning to parent difficult and frustrating. Sitting quietly on the bus was the most peaceful hour I'd spent since his birth.

---

1. Also known as windows. Some of them even opened.

2. They aren't necessarily required to wear them well. No group of officers in business casual clothing is complete without at least one guy wearing his PT sneakers.

When we arrived in Abilene, we pulled into the Eisenhower Museum, put on our Stetsons, and started our tour.[3] As part of the staff ride, as this sort of event is known, each lieutenant had to give a brief presentation on a portion of Eisenhower's life to the group. Some of those presentations were amazing. Mark Ehlers had put together a particularly good roll-up of Eisenhower's early career which was both well-researched and insightful.[4] Others were less good. Lieutenant Milner's presentation, for example, relied heavily on him reading aloud the plaques that happened to be around the spot where he was standing. I'm not sure what Crider got out of our presentations, except that most of his guys were bad dressers who knew very little about history.

After the staff ride, we all had lunch before taking the bus home. One topic that came up in conversation was the Army's newest field manual which was rumored to be coming out soon, *FM 3-24: Counterinsurgency*. In 2006, counterinsurgency was the topic *du jour* for every unit in the army, and since the manual was based on David Galula's *Counterinsurgency Warfare,* Galula's book was on every unit's reading list. The reason for the widespread interest in counterinsurgency was that the war in Iraq wasn't going well, and the Army was trying to adapt its approach to find a way to win.

The initial invasion had been a crushing victory. Spearheaded by the 3rd Infantry Division, coalition forces had smashed Saddam's military with remarkable celerity, and by the end of 2003, Saddam was captured, the Iraqi Army was disbanded, and victory had been declared. Unfortunately, the situation quickly deteriorated, and over the next three years, coalition forces failed to establish stability and governance in Iraq. There were many reasons for this failure, but by 2006 there was widespread recognition in the Army that the approach which had worked so well in defeating Iraq's military—combined arms warfare—was not allowing the coalition to pacify the population, establish governance or rebuild the nation.[5]

Combined arms warfare had taught generations of Army leaders to think of war in terms of the enemy and terrain. The enemy was ahead, friendly areas were behind, and civilians on the battlefield were obstacles to avoid while seeking to destroy the enemy forces. Commanders arrayed their forces to control major highways and infrastructure,

---

3. All cavalry guys are obliged to wear black Stetson hats. Some people like Alex Torres looked pretty good in them. I looked like a dope in a badly fitting cowboy hat.

4. Mark went on to leave the Army and teach history, which surprised exactly zero people in the battalion.

5. Council on Foreign Relations. n.d. *History of the Iraq War.* https://www.cfr.org/timeline/iraq-war.

and established rules of engagement that maximized the distance between coalition forces and the population, to protect both. Unfortunately, this left the population vulnerable to crime and sectarian violence, both of which rose markedly as the Sunni and Shi'a populations of Iraq fought for control of the new democratic government. From the Army's point of view, these were civilian problems, best handled by local police. But as the years passed, political pressure for a U.S. military-led solution grew, and the Army recognized it needed to change. Victory required a fresh approach, and this approach had to be coupled with cultural, organizational, and educational changes in the Army.

The name of this fresh approach was counterinsurgency, and, led by General David Petraeus, the Army invested heavily in changing the mentality of its soldiers to better conduct counterinsurgency operations.[6] Counterinsurgency recognized that controlling the population rather than the terrain was the key to victory. In counterinsurgency, the military's role was to embed itself within, rather than remain apart from, the population. This might sound like a minor distinction, but it was actually profound, and the changes it necessitated were incredibly difficult to implement.

From the first day our soldiers entered the Army, they were enculturated to kill people, not befriend them. Their equipment, embodied by the Abrams tank and the Bradley Fighting Vehicle, was designed for this, and their training reinforced this mentality. Counterinsurgency required units to engage with the population and gain their trust. Not surprisingly, many Army leaders were opposed to doing this because it increased the risk to their soldiers or, in some cases, because they just didn't want to. While Petraeus's campaign to change the Army's mindset ultimately succeeded, leaders across the army had to push hard to make it happen, and even when they did succeed in convincing their units to take counterinsurgency seriously, many still had no idea how to actually implement the approach on the ground.[7]

We fought this ideological battle in the 1-4 CAV about a hundred times that year. Some guys were believers in the counterinsurgency approach. Others, combat veterans

---

6. Counterinsurgency wasn't technically a new approach for the U.S. military, having been employed extensively by the Army and Marines in Vietnam and elsewhere. Association with Vietnam did not increase its appeal, particularly among the most senior leaders, many of whom spent their early years in the Army during Vietnam's aftermath.

7. It took me another fifteen years of military service to recognize the significance of Petraeus's success in transforming the Army's culture. Big organizations don't change quickly or easily, and having failed to achieve less drastic changes in smaller organizations myself, I stand in awe of his achievement.

in particular, were more skeptical or thought it sounded like a suicide mission. As we ate lunch in Abilene that hot July day, the discussion raged again.

"Have you guys read Galula yet? The book is brilliant. His problem in Algeria was almost exactly like the problem we'll be dealing with in Iraq, and he's got some great ideas on how to deal with it."[8] The speaker was Lieutenant Travis Lee, the squadron's tall, thin intelligence officer and one of the smartest guys in the unit. "He says you've got to get out there and really engage with the population to win because ignoring the people allows the insurgents to sway their opinion and use them to hide from security forces."

J.J. Simonsen, the brash, Ranger-tabbed, commander of Bandit Troop, loudly responded from down the table. "That's bullshit, Travis. You can't tell who's who over there. You let those people get close to you, and, sooner or later, you're going to eat a VBIED."[9] Like all our company commanders, Simonsen had redeployed from combat within the last year, and like most GWOT veterans, his experience had taught him that distinguishing friend from foe when the enemy didn't wear a uniform was nearly impossible.

I listened to the exchange, chewing my lunch. I had read Galula but was still inclined to take Simonsen's side of the argument. I had been a machine gunner in Afghanistan and was very familiar with the difficulty Simonsen was describing. On patrol, cars and people had regularly approached our platoon, and it had been impossible to tell whether or not they were going to explode. A few of those explosions and the accompanying ambushes by men in civilian attire had left friends dead or injured, and I was disinclined to give people the benefit of the doubt. The idea that I would live in town, surrounded by Iraqis, seemed suicidal.

Our drive home from Abilene was hot and uneventful. The staff ride had been a welcome diversion from the busy months we'd spent getting the unit ready, and next week we'd be back to work. The troop was coming along well. All the platoon leadership teams were working well together, and the NCOs were whipping the new guys into shape with a heavy training regimen. The discussion we'd just had over lunch led me to consider the content of that training, though. Our NCOs had grown up training

---

8. The book Travis was referring to was David Galula's *Counterinsurgency Warfare: Theory and Practice.* Written by a French officer conducting counterinsurgency in Algeria, *Counterinsurgency Warfare* was a hot book in the U.S. Army in 2006.

9. A VBIED, or vehicle-borne, improvised, explosive device, is a car bomb.

for combined arms warfare, and that's what they were training the guys to do now. Counterinsurgency had percolated down to the platoon leader level, but no further. In part, this was because the NCOs were teaching the guys what they had learned to do themselves, both in training and in combat, but it was also because there weren't any venues or resources to train counterinsurgency skills. Even if we acknowledged that counterinsurgency was the best approach, it looked like we were going to train to do what we knew how to do best: kill people. We'd figure the rest out when we got to Iraq.

We arrived back at the squadron just in time for First Sergeant Strong's safety briefing. As the other Apache Troop officers and I walked up, he was cautioning the soldiers not to "go out there and get shacked up with some stripper with five kids thinking you're going to be her hero and save her from poverty. She ain't in love with you. She's just looking for a ticket for the gravy train, preferably a nice, stupid ticket with E-3 rank on it."[10] The Friday safety brief is a staple of army culture. It's a chance for First Sergeants across the force to put on a bit of a performance, dispense fatherly wisdom, and check a mandatory block, all at the same time. Some First Sergeants view it as a chore and conduct it unenthusiastically, but most of the ones I've worked with have a good time with it and really wax poetic. First Sergeant Strong's were the second best I've seen in the Army, and the troop's laughter punctuated most of the brief.[11] Strong finished up just as the cannon blasted and the bugle sounded "retreat." We all saluted, heard a few words of encouragement from Captain Cook, and headed home for the weekend.

---

11. Uncle Mike's dry, critical, New Jersey-accented tirades are still the best. Alycia accuses me of having Stockholm syndrome for thinking well of him.

5

Camp Merrill, Georgia, November, 2006

I walked past a Ranger lying on his back in the weak November sunshine in Dahlonega, Georgia. He slept on a patch of green grass with a cherubic smile on his face, his hand lodged down the front of his partially unbuttoned pants, and a half-eaten frozen lemon meringue pie on the ground next to him. It was Thanksgiving, and we had just finished the field exercise at the end of Mountain Phase of Ranger School. The exercise had been a miserable ten-day stretch of trudging up and down the mountains of northwest Georgia with minimal food and sleep. We had rescued heavy mannequins named Randy and ambushed convoys of made up-bad guys, and now those members of our class who had passed their evaluations were enjoying a two-day break before flying to Florida for Swamp Phase.[1] Sleep and calorie deprivation make people do weird things, and during the break, Rangers could be found all over the small base, eating the foods they'd fantasized about for the last two weeks, napping, or trying to get their equipment back into clean and serviceable condition. After eating a pile of hot dogs with peanut butter, the foods I'd dreamed about during our last walk through a forty-degree rainstorm, I headed to the long, low row of pay phones to call Alycia.

"Hey, Babe, how's it going?"

"Well. How are you doing, though? Where exactly are you?"

"I'm in Dahlonega, and I'm still healthy enough. I passed my patrols, so I'm going to Florida. Looks like I'll still graduate before Christmas unless I get eaten by an alligator or something."

"That's good, because we just got notified you guys are going to Iraq when you get

---

1. Those who failed would either repeat Mountain Phase or head back to Fort Benning to join The School for Wayward Rangers and wait to start the course over entirely. May God have mercy on their souls.

back."

"Who put that out? I thought everyone was still at NTC? When are we leaving?"[2]

"Soon, I think. They got back from NTC a couple days ago, but they had the FRG handle the deployment notification while the unit was still in California. I'm not sure whose idea that was, but it definitely caused a ruckus."

The FRG was the family readiness group. FRGs exist across the Army to pass information and provide an informal support mechanism for the spouses of service members who need help. Aside from our attached support company, the Raider Squadron was still an all-male unit, so the FRG was comprised entirely of the wives of the married members of the unit. Most of them had been in the Army community for less than a year, almost all of them were living far away from the friends and family they grew up with, and many of them had young children. The FRG served as an *ad hoc* community for these women. It coordinated baby showers for pregnant wives and meal trains for women who had recently given birth, and it served as a social circle. The 1-4 CAV's FRG never suffered from the poor reputation some FRGs had for being hotbeds of dissent, but it was undeniably a rumor mill. Alycia and her friends often knew what the unit was planning, even with no official notification. Using the FRG to notify the unit of an upcoming deployment was a puzzling decision bound to cause trouble, particularly with all the husbands away training.

She continued. "They didn't put out a hard date yet, but I think you're going to be leaving right after the holidays."

"Shit. Iraq probably isn't the best place to recover from Ranger School."

"No, probably not. I don't think you're going to be home for even a month."

"Ugh. I'm not sure I'm going to be able to walk right after this course, much less fight a war. I guess that's tomorrow's problem, though. Right now, I need to just get out of here!"

"Well, good luck finishing up. I can't wait for you to get home! I'll bake you something amazing."

"That sounds great. I can't wait!"

We wrapped up the phone call after a few minutes of catching up, and I hung up the phone. I sat down for a minute and thought through the situation. We were

---

2. NTC is the National Training Center in California. It's a giant desert training area where units go to get certified by the Army to deploy to combat.

headed to Iraq. RUMINT suggested the Army had increased the standard length of combat deployments from six months to a year, and having just come off a nine month deployment in 2004, I knew this one would be tough on the family. Before I could deal with any of that, though, I still had another two weeks of Ranger School to get through.

The rest of Ranger school passed in a blur. By that point in the course, we had been deprived of food and sleep for so long that the conscious portions of our brains had mostly turned off, and we staggered through the swamps of Florida on autopilot. We piloted boats down rivers, waded through hip-deep water with 80-pound rucksacks, and assaulted beaches, but I remember surprisingly few details. My body started to break down about a week into the final field problem, and my arms and legs swelled with Staph infections until I could no longer take my pants off or roll up my shirtsleeves, but I made it to the end of the course with no serious mishaps. I even managed to graduate as the top guy in the class, a laurel I credit entirely to the blessings of the weather gods.[3] At Crider's insistence (and with a large amount of his help), Alycia came to the graduation ceremony with Gareth to pin on my Ranger tab, and after a hearty lunch of Mexican food, we flew home to Kansas together.

Through the rest of December and January, rumors swirled related to the information Alycia had given me on the upcoming deployment. According to one rumor, the squadron was heading to the southern part of Iraq to secure convoys traveling north from Kuwait. Another implied we were being instead tasked to support an operation in Sinai, Egypt. A third stated the deployment was going to be canceled entirely and insisted the squadron was only being alerted as an internal exercise. As it turned out, though, the information Alycia had given me on the phone in Ranger School was the most accurate, and the squadron was scheduled to deploy to Kuwait in February and to move north into Iraq from there.

The last month of preparation was busy at work and stressful at home. At work, there were a thousand things to do. The squadron had returned home from the National Training Center in California right before Thanksgiving, and equipment had to be cleaned and repaired. New wartime communication equipment had arrived, and the operators who would use it had to be trained on its use. Soldiers had to be trained and

---

3. I have a hypothesis that whether a given student passes Ranger School is strongly correlated with the weather during his graded patrols. Rangers are evaluated on their ability to lead missions, and success depends largely on how well their fellow students perform. Bad weather makes Rangers dumb, though, so even the best leaders are likely to fail if it happens to be particularly cold, wet, or miserable on the mission.

certified on how to load equipment on the trucks and train cars which would take it from Kansas to the east coast shipping ports where it would embark on the trans-Atlantic voyage to Kuwait. Finally, everyone in the unit needed to take online training and get briefed on a seemingly infinite array of topics, ranging from Arab culture, to the conduct of military operations in the desert, to counterinsurgency best-practices.4

Home life was similarly busy. Figuring out what you need to take for a year-long trip is difficult since it's not as though you can pop back to the house for a second to grab anything you might have forgotten. There were duffle bags to pack, last-minute trips to the sew shop to take so equipment could be repaired or labeled, and lost pieces of gear to either find or replace. All of this was difficult, albeit familiar, churn that Alycia and I were both accustomed to. This time, though, something was different, and the deployment preparation process was unusually unpleasant.

Alycia and I had been married for two previous deployments, so we understood the prospect of death was part of wartime service, but the last two deployments had come suddenly, with little time for preparation. The long, reasonably predictable pre-deployment period changed the dynamic substantially. The upcoming deployment gave all the normal home activities a dark undertone. Was this the only Christmas we'd have together as a family? We weren't sure, so we took more pictures than usual and tried to capture the experience more thoroughly. If I died, would Gareth remember any of the time we'd spent together? We were pretty sure he wouldn't, and even if I lived, he probably wouldn't remember me when I got home, so we recorded a dozen videos of me reading his favorite books and ordered a stuffed Daddy Doll for him.5

We also had to make more practical preparations that were just as difficult. First, I had to get a death photo produced. It's just as macabre as it sounds. The death photo is an official photograph taken of every deploying service member with an American flag in the background. The photo is kept on record so that if the service member is killed in action, the Department of Defense can attach it to the official death notification. We

---

4. Some of these classes turned out to be useful, like the information presented on the different sorts of improvised explosives being used in Iraq at that time. Others were less so, like the briefs on Arab culture. I spent the first six weeks of the deployment awkwardly trying to ensure I never showed the soles of my feet to anyone before my interpreter let me know it wasn't really that big a deal.

5. The Daddy Doll was a small, plush kid's toy with a silk-screened photo of me on it. Daddy Dolls were produced by hugahero.com during the GWOT to ensure kids didn't forget what their deployed parents looked like. Gareth loved his.

also had to update our life insurance information. We needed more coverage than the standard policy the Army paid surviving spouses, so I took out an additional policy. Life insurance physicals are a normal part of life, but taking one because you expect to die feels strange. These proceedings, and our discussions about who would look after Gareth if we both died, what the funeral arrangements would look like if I was killed, and what Alycia and Gareth would do without me, cast a grim shadow over the predeployment period.

On February 4th, the preparations finally came to an end, and the squadron assembled in the parking lot behind the headquarters to board the buses to the airfield. Over the last few months, the squadron membership had changed a bit. Crider had been individually assessing leaders across the squadron, and between those assessments, various people's performance at NTC, and the natural ebb and flow of the Army personnel system, some changes had been made. Moon was too senior to remain in his position and–despite his protestations and Crider's recommendations to higher–had to leave the squadron.[6] Several other captains, platoon leaders, and sergeants were rotated out as well, and their replacements were in the parking lot getting the squadron ready to deploy. Even Jeff, my old friend from basic training, was pulled off the deployment roster. The scars on his legs weren't the only wounds he'd suffered in Iraq, and over the course of the year, his intensity and kill-em-all mentality had convinced his command he needed to stay back in Kansas to cool off a bit. He hadn't taken the news well, and he was in the parking lot now, hanging out around the edge of the crowd in disbelief that he wouldn't be able to deploy with the "cherry-ass fucks" he'd spent the year getting ready for war.

The going away ceremony had been arranged by the squadron's new rear detachment commander, the former Comanche Company commander, Tom Laybourne. He was about to pin major and too senior in rank to command the troop in Iraq, so Crider tapped him to manage the squadron's rear detachment while the rest of the unit was forward. Being selected to serve on the rear detachment, or rear-d, was a job nobody wanted. The rear-d passed information to the spouses of deployed personnel, coordinated replacement soldiers' movement forward into theater, and arranged the funerals and medical treatment of returning personnel. Since rear-d personnel were both the bearers of bad news and still at home with their families, they frequently served as a focal point

---

6. For all the (justified) pain he inflicted on the company XOs, MAJ Moon loved the battalion, and he took this transition hard. He was replaced as Battalion XO by Major Tim Baer.

for spousal rage and frustration. Although it was unfair, rear-d personnel were often viewed with a combination of contempt and envy by both the deployed soldiers and their families, and in return for performing this thankless job, the rear-d received few accolades.

The ceremony itself was painfully awkward. The weather that day was cold and windy, and the mood in the parking lot was bleak. A long row of empty buses that would take the squadron to the airfield dominated the street and lurked in everyone's peripheral vision as the soldiers of the Raider Squadron tried to say goodbye to their families. Soldiers with their crying wives and children huddled in small groups around the cold parking lot, and leaders tried to balance saying their own goodbyes with their responsibility to get their personnel on the buses in an organized and timely manner.

In the middle of all this, Alycia, Gareth, and I tried to say our own farewells.

"So, I guess this is goodbye."

"Yeah, I guess so. Take care of yourself. Gareth and I are going to miss you."

"I'll miss you too." I checked my watch and looked around to see where Captain Cook and First Sergeant Strong were standing. I knew we had fifteen minutes to get everyone on the buses, and that it was going to be difficult to find all the Apache Troop guys scattered across the parking lot and pull them away from their families. Alycia noticed my distraction, and I could tell it hurt her feelings. I wouldn't see Alycia and Gareth for months, and while I knew I should focus on my own family, I couldn't stop thinking about all the things I had to do to get the troop out the door. What was wrong with me? I felt guilty and conflicted until Alycia interrupted my split attention.

"Just go. You're distracted."

I felt bad, but she was right. Nothing good would come from her and Gareth being here. It only highlighted the fact that I was more focused on getting the troop to Iraq than I was on leaving my family for a year.

"I don't want it to be like this. I'm going to miss you both terribly. You're right, though, I'm distracted. I have to get the guys on the bus, or we'll blow all our deployment timelines. I love you, and I'll call you as soon as I get somewhere with a phone."

Alycia nodded. I kissed her and Gareth, and we walked over to the bus. I patted Gareth's head and hugged and kissed Alycia again. We looked at each other for a moment sadly, then I checked my watch and walked to the bus. First Sergeant Strong was already there, and the two of us herded the men of Apache Troop onto the buses. As the last guys climbed on, Strong went to find Cook for a last huddle with the squadron leadership.

I looked out of the window at the parking lot full of crying families. Behind them all, I noticed Jeff. He was standing next to his pickup truck with his arms crossed, with an expression which conveyed the anger, embarrassment, and confusion he obviously felt about being left behind as everyone else went to war.

"Ah well," I thought. "The Army goes rolling along."

As the buses drove off, I looked the other way toward where Alycia and Gareth were still standing on the sidewalk near my window. Gareth waved uncomprehendingly, and Alycia smiled with tears in her eyes. I smiled back sadly and waved at them as we pulled away, wondering if I'd ever see my family again.

Part II

Combat XO

Grey Zone Ethics LLC

6

Camp Buehring, Kuwait, February, 2007

"While you're in theater, you are not authorized to consume alcohol of any kind or to gamble. You also can't possess or look at porn, and you can't have sex with other service members or local nationals. Don't enter any mosques or religious sites, and don't discuss religion with the locals."

I glanced up momentarily from the lawyer's class, and my eyes were greeted with the unwelcome sight of Lieutenant Jarvison sitting spread-eagled on his cot and rubbing a large amount of white ointment onto his exposed crotch. The men of Apache Troop were living in a large, thirty-man tent in Kuwait while they waited for an aircraft to take them to Iraq, and to fill the time until then, the company was receiving a class from the JAG on the rules we had to follow while we were deployed. His makeshift classroom was set up between the two rows of cots that lined the tent's walls. Lieutenant Jarvison's cot was located directly behind the lawyer's instructional white board, and unfortunately, Jarvison had somehow failed to notice the twenty soldiers looking attentively in his general direction when he'd decided to administer a bit of self-care for his heat rash. The crowd's snickering eventually got the instructor's attention. He turned to look at what was distracting the crowd, stared open mouthed for a second, then shook his head and continued the class. Lieutenant Jarvison was one of the fire support officers in the squadron, and his strange antics were well known.[1]

As I continued watching the presentation, a sudden ripple of discomfort hit my gut that indicated nature was calling, so I stood up and left to resolve the issue. I stepped outside the tent and was immediately blasted in the face by the hairdryer-like heat of

---

1. Lieutenant Jarvison was also an enthusiastic nunchucker. He'd brought a set to Iraq and put on demonstrations for the soldiers sometimes.

daytime Kuwait.[2] I grimaced and started the half-mile walk to the Porta-potties. Camp Buehring, Kuwait was an enormous halfway house for units rotating into or out of theater. The process of moving everyone from home station, waiting for the equipment sent from Kansas to arrive by ship, issuing theater specific equipment, and conducting the final pre-combat training took a few weeks, and Buehring was where units waited while it happened.

Buehring had that special neglected feel that all the unloved, transient places of the world have. The camp was largely composed of tents, connexes, and prefabricated buildings interspersed with fields of gravel and dust, and connected by wide, hot sidewalks. Vast banks of graffiti-adorned porta-potties squatted, reeking in the sunshine, and aircraft pallets stacked high with sagging mountains of tepid water bottles and labeled with "hydrate or die!" signs were strewn about haphazardly.[3] The air constantly rang with the sound of high-powered aircraft engines, and the smell of jet fuel, chemical toilets, and exhaust filled the air.

The occupants of this weird place looked similarly unloved and transient. Almost all of them were either on their way into combat or on their way back home from it. Most were disheveled and uncomfortable as their circadian rhythms struggled to overcome time zone changes and their sinuses wrestled with the combination of dry, desert dust and frigid tent air conditioning. During the day, they shuffled to and from the islands of climate control, and at night, they lounged outside talking or playing pick-up basketball. If the Catholic idea of purgatory is a real thing, I imagine it looks like Camp Buehring.

The walk to the nearest bank of Porta-potties took about five minutes. I found one with the small green disk above the handle that showed it was unoccupied, took a deep breath of the relatively fresh Buehring air, and stepped into the small plastic cell. It was late morning, so the internal temperature of the Porta-potty was only about 110 degrees, but despite my preparation, the mixed aroma of urine, feces and chemical sanitizer assaulted my nose like a clenched fist. Before I even got my pants down, my face was beaded with sweat, and by the time I had untangled my slung weapon, found a place to hang it, and gotten myself seated, I was on the cusp of blacking out from heat exhaustion.

There's an art to using a Porta-potty. If you do it incorrectly, you get a backsplash

---

2. According to weatherspark.com, the day-time temperature in Kuwait is rarely above 122 degrees Fahrenheit and usually sits around a more comfortable 115 to 117 degrees.

3. The Buehring porta-potty graffiti was incredible. I've always hoped someone would capture its magic in a coffee table book.

of blue water on your ass, and I'm not sure the medical community has any cure for what you get when that happens. As I tried to carefully work through this process, I was interrupted by a strange series of sounds from one of the Porta-potties to my right.

"Mmm...Ahh...Ugh...ooo."

My sphincter froze. What was I hearing? The sounds continued, and they were punctuated by occasional giggles and the banging noise of something thumping against the Porta-potty's plastic wall. "Oh God," I thought. "There are two people screwing in one of the other Porta-potties."

As the realization of what was occurring dawned on me, I faced a dilemma. Should I finish quickly and try to leave before they were done, or should I sit in this reeking hell until they finished? In either case, I definitely didn't want to leave at the same time they did. Screwing other soldiers was a punishable offense in theater, and while I couldn't have cared less about the infraction, as an officer I was obligated to act if I recognized either of the soldiers involved. That meant I'd have to fill out a ton of paperwork *and* that I'd ruin a couple of soldiers' months, because they'd likely lose pay for the infraction. It'd be even worse if one of them was married.

As the sweat poured down my face in the overwhelming heat, and the rhythmic thumping and moaning continued next door, I decided a speedy exfiltration was the best plan and hastily finished my business, gambling that I could get out of the area before the enamored couple next door finished theirs. I buckled my belt, grabbed my weapon, and threw open the plastic door. The blast of relatively cool, clean air felt miraculous. Not pausing to savor the sensation, I stepped hastily out of the cell, put on my hat, and strode to my tent without looking back.

The lengths people sometimes go to get their rocks off is amazing. The experience of taking a dump in Kuwait was one of the most unpleasant experiences I've ever had, and I was alone and conducting a relatively stationary activity. I can't imagine adding another person and a few minutes of strenuous physical exercise to the situation.

When I got back to the tent, the lawyer was still teaching his class, so I sat down on my cot to listen to the last hour while I tried to wash the taste of the Porta-potty out of my mouth with a bottle of water and forget the last ten minutes of my life. When the class was finished, the troop dispersed to work out, make phone calls, or listen to music. I went to lunch.

The walk to the chow hall took about ten minutes. On the way, I bumped into Rob Humphrey and Alex Torres, who were leaving the post exchange and heading to the same chow hall I was.[4] I joined them and asked how things were going.

"Hey y'all. I just left the tents where the 2nd Platoon guys were getting their JAG briefs. Everyone seemed super excited to get a refresher on the Law of War."

Torres responded, "Yeah, Kuwait sucks. The guys are just sitting around, and they're hot and bored as hell. When are we heading to the range? Maybe some shooting will help them blow off some steam."

I nodded in agreement. "Y'all find anything good at the PX?

"It was pretty picked over, but I grabbed a copy of *Talladega Nights* and a Coke."

Retail therapy was a common way for guys to cope with the stress of deployed life. Some guys ended up putting together extensive home entertainment systems in their tents, containerized housing units, or repurposed Iraqi barracks rooms, and almost everyone with PX access dropped in for at least a periodic bag of beef jerky or a log of dip.[5]

"Sounds like you've got quite the party planned tonight, Alex."

"Ha! Yeah, I guess so. All I need is a few *chicas* and *maragritas.*"

My thoughts drifted momentarily back to my Porta-potty experience. I shuddered. Maybe Alex should stick with just the *maragritas.*

We finished the walk to the chow hall, washed our hands at the foot-pump powered sinks outside, and entered the chow hall through the airlock the staff maintained to keep dust out of the building. The contrast between the sweltering conditions outside and the cool, pleasant interior of the chow wall was stark. As we grabbed our trays and walked through the buffet line, I marveled at the facility.

One side of the room offered regionally themed dishes from a half dozen countries. The smell of enchiladas, sweet and sour chicken, a steak bar, and stir-fry tempted my nose. In the center of the room, a spectacular salad bar offered all the healthy (and unhealthy) toppings a person concerned with his weight or vitamin intake could ever want. Finally, on the opposite side of the room sat the dessert bar, where ice cream, Otis

---

4. The PX, or Post Exchange, is an on-base store that imports and sells American goods to service members. The Exchanges in Iraq and Kuwait sold decent pillows and blankets, CDs and DVDs, and electronic equipment. The only alternative, at least for guys who left base to patrol, was to buy stuff locally, which had its own problems, like finding out you'd accidentally bought a cheap knock-off or getting blown up by an IED.

5. Copenhagen if a guy was into the good stuff. Grizzly if he was cut from thriftier cloth. Dip was a valuable currency everywhere except the huge FOBs where it was easy to come by, so guys tended to stock up. Once guys ran out of dip, addiction ensured they were willing to pay top dollar for a can.

Spunkmeyer muffins, and an under-used fruit section fought to ensure every deployed soldier increased his body mass on the rotation.

One of the most incredible achievements of the GWOT was the perfection of expeditionary consumer logistics. On my last trip to Afghanistan in 2003, post exchanges in our remote, mountain FOB were non-existent, and chow was either locally procured Afghan food or prepackaged military rations. We lived in a twenty-six-man tent with a single gas stove in the center and no air conditioning, and we slept on cots. Our showers were cold, short, and infrequent, and we burned our own feces in open, round cans full of fuel that sat under homemade outhouses. Communication was similarly austere. Our base had a single satellite phone. Each guy was allotted a ten-minute phone call every two weeks, and even that rarely worked.

From 2003 to 2007, the Army had perfected its ability to provide deployed soldiers with creature comforts, so by the time we deployed to Iraq, a combat soldier's standard of living was often significantly different. When on a major FOB, a soldier could expect an air-conditioned room with electricity and a hot shower down the hall. Banks of AT&T phones and internet service provided communication with home, and chow was so good—every Friday, steak, shrimp, and lobster tails were served—that people often put on weight while deployed.[6] Life was different for those of us who spent our deployments on smaller outposts, but even we enjoyed trickle-down benefits like Ripits from the logistical marvel that was FOB life in Iraq. These creature comforts had unforeseen consequences that I discovered later, but in Kuwait, it seemed pretty incredible.[7]

As I munched on a piece of shrimp, I wondered aloud to Alex and Rob, "What do you think it costs to get every guy in Iraq a lobster tail every Friday?'

Rob thought for a minute and replied, "No idea, but at least there's a port here. Imagine how much it costs to get this stuff into Afghanistan. Everything there has to be shipped by truck through the mountain passes of Pakistan. I'm sure the tribes there have to be paid off, and even when it gets to a big FOB, it's got to be flown to the outstations by helicopter. I bet a lobster tail in eastern Afghanistan easily costs a thousand bucks."

Alex replied, "I wish they'd keep the lobster and give me the cash."

We all laughed and continued eating, looking up occasionally at a nearby television

---

6. That had not been the case in Afghanistan. Thanks to a variety of colorful digestive problems collected from the local cuisine, most of my guys had lost a good bit of weight on our deployment.

7. The availability of Ripits turned out to be a mixed blessing. Several of our guys ended up getting kidney stones on the trip because of overconsumption of the popular energy drink.

which was playing the Armed Forces Network's (AFN) hourly news program. On it, an impeccably groomed Air Force Airman described the day's current events and provided a weather forecast for the deployed areas of the world. Sure enough, Kuwait was going to remain warm for the foreseeable future. When the forecast was complete, a helpful AFN commercial reminded us to never shake a baby. Even supposing I didn't know that, I was in Kuwait, so my odds of shaking my baby were negligible. My baby was safe. Thanks, AFN.

When we finished, I stood up with Alex and Rob to turn in our trays to the courteous Indonesian dish guy and head back to the squadron area. Alex and Rob split off to find their platoons, and I walked back to the troop headquarters to find Captain Cook and First Sergeant Strong. I found them sitting at a folding table next to an easel with a large pad of paper hung on it. Cook had a marker, and he and Strong were talking through the rest of the tasks necessary to finish our time in Kuwait. Privates Russel and Banninger, the commander and First Sergeant's drivers, were seated nearby and fiddling with a radio to get it working. They were quiet, organized guys who had been singled out early as the sort Strong needed to run a functional office. Their ability to keep track of tasks and paperwork was remarkable, and they were largely responsible for ensuring our troop functioned on a day-to-day basis.

When I walked in, Cook turned to me. "XO! Where have you been fucking off for the last hour? Did you line up the ammo for tomorrow's range?"

I took my seat, wiped the sweat off my face and popped open a bottle of water. "Yes. Sir, it's all lined up. We've got rifle ammo for all three platoons and machine gun ammo for the gunners. There's enough to zero and qualify everyone and to conduct the maneuver training you wanted to get done."

Cook nodded and grinned. "Good, we need to get the guys doing something other than watch movies and sit on their cots. Nothing good comes out of a bunch of idle troopers."

Strong laughed. "Yeah, we've already had soldiers from the support battalion get caught railing each other in the Porta-potties, and it's only a matter of time before one of our guys figures out how to get in on that action. I don't need that kind of paperwork in my life."

I nodded in reply, keeping the most neutral expression on my face I could manage. *I* didn't need that kind of paperwork in my life, either.

The next topic of conversation was trying to figure out where we'd be going in Iraq.

Surprisingly, there was still uncertainty about this, but the most likely plan was that the squadron would fly into FOB Liberty, a base in the Green Zone area of Baghdad, pick up our vehicles, and drive the short distance south to FOB Falcon, where we would be based.[8] From a company XO's point of view, the details of getting there weren't too difficult. Now that we were in the big, green, deployment machine, we just had to wait while the bureaucratic and logistic wheels turned until we came out the other end.

The rest of the meeting was a routine review of the various statistics we had to report to Squadron. The Troop had such-and-such percent of our guys in Kuwait so far. This many guys had completed their theater-required medical training, and that many guys still needed to draw their individual medical kits from the warehouse. We finished our review, ended the *ad hoc* meeting, and I left to find the Troop's communication sergeant, Staff Sergeant Bonilla, to drive out to the range and make sure everything was ready for the next day's marksmanship training.

Bonilla and I signed out an armored SUV, drove off the base, and took the adjacent highway toward the ranges. Beuhring is in the middle of nowhere, so we had to cross miles of desert to reach the range complex. As we drove down the highway, we looked out across the dunes of the Kuwaiti desert and marveled at what we saw.

"Dude, look, it's a no kidding camel caravan."

Sure enough, it was. In the shimmering, hazy distance, a line of camels tethered together end-to-end was walking slowly across a dune. In front of them, we could just discern the tiny figure of a person leading the caravan to God knows where.

"Wow, that's like something out of the *Arabian Nights*."

As we took in the sight, our SUV passed a wrecked Bentley. The ruined car was on its side in the ditch that ran alongside the road, and the car looked like it had been there for a long time. That day, the juxtaposition of the wrecked luxury car and the nomadic caravan was a strange sight, but over the year I got so accustomed to the bizarre contrast between the wealthy, modern society present in much of Iraq and Kuwait and the medieval poverty of the tribes which existed around the edges of Arab culture, that it stopped being remarkable. The Bedouins who roamed across and camped in the open desert in Kuwait, the tribal shepherds who drove their flocks of spray-painted sheep

---

8. The Green Zone, colloquially known as the Emerald City, was a large, secured area in the center of Baghdad where the theater headquarters, embassy, and government of Iraq were located. With its luxurious accommodations, restaurants, and noticeable lack of IEDs, it looked idyllic to the people who spent their deployment in the more austere parts of Iraq.

through the urban center of Baghdad, and the groups of masked women wrapped from head to toe in colorful cloth who sifted through dumps for food for their goats, just became part of the colorful background of our wartime experience.[9]

Looking at the nomads, I remarked to Bonilla, "Maybe that's true libertarian paradise?"

Bonilla laughed. "Maybe so, sir. Think they'd be willing to hang a 'Vote Gary Johnson' sign on one of those camels?"

"Ha! Yeah, maybe so."

We continued driving, zigzagging our way through the Kuwaiti training areas until we found the range we were seeking. It was behind a locked gate, but the number on the gate matched the one we were looking for, and everything else looked in order. "I guess we can't get in, but at least we know how to get here now. Let's head back and tell the boss we found the range. They'll figure the rest out tomorrow."

---

9. These women's attire, affinity for dumpster diving, and ability to sneak into almost anywhere earned them the moniker of "trash ninjas." They never spoke to us, and they never caused us any trouble. The Iraqi locals viewed them with scorn, but never impeded their trash collection.

7

FOB Falcon, Iraq, March, 2007

The night was cool and dry. The massive flood lights arranged around FOB Falcon, the squadron's new home in the Doura region of Baghdad, bathed the entire compound in bright, white light to prevent enemy forces from using shadows to sneak over the concrete walls of the base. Usually, this was a good thing, but tonight, the bright lights were our enemy because they kept Staff Sergeant Curtwright, Sergeant Enfield, and me from staying out of sight as we stealthily made our way across the base on our secret mission. Our mission that night was to deface a hated emblem of a conflicting ideology that was brazenly displayed within eyesight of our own squadron's headquarters. Apache Troop had long been aware of the emblem's existence, but thus far, it had been unassailable. Tonight, though, by an audacious commando raid we'd dubbed *Gato Negro*, it would fall.

Our preparation for the mission was compartmentalized–even our own commander didn't know about it–and in the vein of commando missions throughout history, the selected force was a small, all-volunteer element, purpose-built for the task. We carefully reconnoitered the target after dinner, moving with the usual flow of on-base traffic to avoid any undue attention, and just after midnight, we gathered at the troop headquarters to conduct our final preparations.

"Alright," I said to my fellow conspirators. "Do we have everything?"

Curtwright nodded and held up his can of red paint. Enfield nodded as well, gesturing toward his can of white. Both had brushes in hand.

"You both have your masks?"

They did. I patted my cargo pocket to ensure my own was still there.

"Let's do this."

The three of us left the troop and began quietly walking across the FOB toward the

1st Battalion, 28 Infantry Regiment's headquarters.

The target of our operation was the black lion sitting outside the 1-28 Infantry's headquarters. The 1-28's moniker was "The Black Lions," and when the unit was stood up the year prior, the battalion commissioned a seven-foot-tall metal lion to post outside the battalion headquarters. For the last year at Fort Riley, that lion had glared at the cavalrymen of the 1-4 CAV as we ran past it during morning PT, drove up Custer Hill to go to work, or walked to the brigade chow hall, and now the battalion had brought it all the way to Iraq to continue staring at us. To the cavalry soldiers of Apache Troop, this audacious display of infantry hubris was a total affront, not to be tolerated. The black lion needed to be humbled. The black lion needed to be calvary-ized. The black lion needed to be painted red and white.

Our infiltration of the 1-28's headquarters area went smoothly, as few people were awake this late at night. We paused about twenty yards from the headquarters and conducted our final planning huddle.

"Alright, let's review the plan. I'll post by the corner of that wall outside the headquarters and distract anyone who comes out of the door. Curtwright will start at the top left of the lion with the red paint and work his way down. Enfield will start from the bottom right with the white paint and work his way up. Once we're done, I'll snap a quick photo, and we walk off nonchalantly and ditch the paint and brushes in the dumpster. We're not looking to paint a Picasso here, but the coverage needs to be good enough to look like the cavalry flag, otherwise it'll just look dumb, and nobody will get the point."

Curtwright and Enfield nodded, smiling. They both pulled the ski masks out of their pockets and put them on. We were wearing PT uniforms, so with our faces covered, we were indistinguishable from anyone else on the base. This was critical. If the mission went south, we didn't want any bad publicity for the squadron or company leadership.

"Good luck, gentlemen, and Godspeed." With those final words, we moved into execution. The operation started smoothly. As the two NCOs began quickly painting the diagonal red and white cavalry flag onto the lion, I watched the door to the headquarters closely for any sign of activity. The entire painting process took about sixty seconds, and as Curtwright and Enfield finished, they stepped back across the street so we could get the photograph. I pulled out the camera and snapped the picture. No luck. The flash hadn't fired, and the picture on the digital camera's small screen was too dark to see. I reset the camera and prepared for another snap, but as I did so, the door to the headquarters opened and a 1-28 soldier stepped out to smoke a cigarette. He didn't immediately look our way, but as the flash on my camera fired, it drew his attention our way, and he looked directly at the dripping red and white lion. In the split second before we ran, I saw something like recognition of a problem pass over his open-mouthed face, but the next second, we were gone, running across the FOB, and I never saw what his reaction turned into. The three of us ran a circuitous path which looped us in the opposite direction from the 1-4 CAV's barracks until we made it back to our troop headquarters, stopping only to dump our paint and brushes in a trash can.

When we got back to the troop area, I pulled out the camera, and we crowded around the display to check the last photo. The flash had worked, and the picture had come out perfectly. Mission complete.

Shaking his head and laughing, Curtwright loudly asked, "Did you see the look on that guy's face? He was like 'What the hell? Oh shit! The lion is red and white!'"

I laughed and replied, "Yeah, you guys crushed it. That paint job was a work of art, and you did it fast as hell."

We all nodded silently in agreement, catching our breath and basking in the glory of a successful mission. The group dispersed, and we all left to get ready for bed.

The 1-4 CAV had been at FOB Falcon for about a week. There at been a brief period on our way into theater when it had looked like we'd be assigned to the 1st Cavalry Division in central Baghdad, but after some last-minute unit shuffling, the theater command had assigned us to southern Baghdad instead. We'd settled into our barracks building–an old Iraqi army barracks–and set up our Troop headquarters on the top floor so the communications team could run all the antennae for our radios onto the roof. The platoons had started their initial area familiarization patrols in *mahala* 826, a.k.a. *Al-Mekanek*, a.k.a. Mechanics, the neighborhood just east of FOB Falcon.

The process of assuming responsibility for an area of operations took about a week. During that period, at the platoon level, guys had to determine how they would set up their trucks' communication, navigation, and weapons systems. They had trained on patrolling, but the actual process of doing so in combat was a bit different and working out the kinks in their internal systems took time. How would the platoon arrange its HMMWV's when they stopped? Who would get out when it was necessary to dismount, and what would the rest of the crew do while the dismounted element was conducting its task? How would the platoon handle traffic? Would they allow cars to pass them or weave in and around the platoon's vehicles, or would they obstruct traffic entirely? And those were just the platoon-level problems. At the troop and squadron headquarters, similar problems had to be solved. How would the squadron react to a platoon hitting an IED?[1] Would the squadron surge another unit to assist the struck element? If so, where would it come from, and how would notification of the response element occur? If there was a casualty, what were the theater processes in place to treat his injuries? What if he died? These and a hundred more questions had to be answered before the squadron commander could send notification to the theater headquarters that the squadron had officially assumed responsibility for the Doura province of East Rashid.

Since many of the issues identified were at the troop and squadron levels, and because the platoons couldn't assume a full patrol schedule until everything was resolved, the guys in Apache Troop had a bit of free time. They had just deployed to combat, and adrenaline was running high. Guys checked and rechecked their gear and burned off excess energy during extra-long workouts at the gym, but as Captain Cook had noted in Kuwait, idle hands sometimes found other ways to stay busy, and the commando mission Curtwright, Enfield and I conducted that night in March was one of them.

I heard nothing about the aftermath of the lion incident, and I didn't inquire. Years later, Lieutenant Colonel Crider told me that the 1-28 commander had taken the whole affair in good humor. At the time, though, I wasn't sure how pissed off the squadron leadership might be over our jack-assery, so I kept my involvement in the incident quiet.

Over the next few weeks, as we began conducting combat operations and people from both units started getting injured and killed, *Gato Negro* looked ridiculous–like Apthorpe's infamous row with Ritchie-Hook over the thunder box in *Men at Arms*–but that night, as I walked out to the showers to clean up for bed, it seemed glorious.

---

1. Improvised Explosive Devices, or IEDs, were a fixture of our effort in Iraq. These homemade bombs, which ranged from the very simple–a coke bottle full of explosive powder with a simple detonator connected to a cell phone–to the very complex–carefully manufactured shaped charges activated by passive infrared sensors that could cut through the armor of a tank–evened the odds between allied forces and the local insurgents.

8

Mahala 826, Iraq, April, 2007

The night was sweltering and hazy, and the lens on my monocular night vision goggles (NVGs) kept fogging up as I picked my way around craters in the street with 2nd Platoon. Under night vision, all liquid looks the same, so it was impossible to tell whether a pool in the street was an inch-deep puddle of urine and motor oil or a three-foot-deep crater of sewage. I assumed the worst and avoided them all. The street was narrow by American standards and in poor repair. There were sidewalks running along the edges of it, but these had been obstructed by abandoned cars and shanties, so we made our way down the middle of the road. On either side of us were battered, one-to-three-story buildings. Most of them were abandoned or entirely reduced to rubble, but occasionally a glowing window threw a patch of light into our path which indicated there were still inhabitants in this urban wasteland.

Our patrol was split into two elements. In the lead, a squad of dismounted guys cleared the road and looked for trouble. Behind them, a pair of trucks with heavy turret-mounted machine guns provided fire support. The headlights of the trucks were turned off to avoid washing out our night vision. We weren't a stealthy formation, since the truck's diesel engines made it clear to anyone listening that we were in the area, but with no visible light, it was difficult for anyone trying to attack us to pinpoint our location. I was walking with Lieutenant Sisoura in the dismounted element, and we'd been walking for about an hour when the point man called a halt on the radio. I took a knee and scanned the buildings around us while Sisoura crossed the street and knelt to scan the opposite direction.

After clearing the space in front of me, I stood up and walked over to Sisoura. He had a platoon radio, so I asked what was up.

"The point man found a dead body in the street. We're trying to figure out what to do with it."

"Need anything from the Troop HQ?"

"No, I don't think so. We'll work it."

I nodded. Nobody in charge likes an extra person adding to his confusion by badgering him with questions, so I scanned the nearby buildings for enemy activity and waited quietly for the situation to develop a bit.

After a few minutes, the truck behind us turned on its headlights. Blinking in the sudden brightness, I turned off my NVGs. Koky turned to me and said, "We're going to white light the body to make sure it's not rigged with an IED, then get the police to pick it up."

"Gotcha, I'll call it up." I called the report up to the troop headquarters and walked with Koky as we guided the truck up the street.[1] After a few hundred yards, the headlights illuminated a lump of something laying in the street with a 2nd Platoon soldier standing over it. As we got closer, the soldier turned out to be Staff Sergeant Enfield, and the lump was revealed to be the bleeding body of an Iraqi man wearing a traditional *dishdash* and sandals. He was bleeding from a fresh gunshot wound to the head; it looked like he'd been executed in the street. As 2nd Platoon spread out to maintain security around the area, Koky and I approached Enfield. He finished up the radio call he was making, then turned to face Koky and me.

"Sir, the Iraqi police checkpoint is telling our guys they won't drive over here to pick this body up."

"Why not?"

"The police aren't being too clear about their reason. They're saying something about having orders not to leave the checkpoint, but I think they're full of shit. They're just scared to come out."

Koky and I nodded in agreement. Enfield was probably right, but it didn't help us deal with the situation at hand. If the police wouldn't come pick up the body, we could address it with their leadership the next day, but in the interim, we had to decide what to do with the body. In a few more months, after we'd found a hundred more bodies just like this one, we would drive over to the police checkpoint and insist they pick up the body themselves, but this early in the deployment, the process was still uncertain.

---

1. Probably the only real value I provided to the platoons on these patrols, other than another rifle if we started shooting, was taking the burden of reporting to the troop off the leadership's shoulders.

Enfield continued, "What do you want us to do with the body? I say we just leave him here. The locals can clean him up tomorrow."

Koky replied, "No, I don't want to just leave him here. We're trying to get this place back to normal, and when people wake up to dead bodies in the street, that's not normal. We need to do something with him."

I thought for a minute, then offered, "We could just throw him on the hood of the truck and take him to the checkpoint ourselves."

The group thought about the proposal, then decided it would work. I took the dead guy's arms, Enfield took his legs, and Koky helped guide the body onto the hood of the truck. After some wrangling, we managed to position him well enough that he'd stay on for the drive to the police checkpoint. Enfield signaled to the driver to start rolling, and we started leading the patrol along the half-mile route to the checkpoint, one of us keeping a hand on the dead guy to make sure he didn't fall off.

The movement passed uneventfully, and we made it to the checkpoint in a few minutes. When we got there, the rest of the 2nd Platoon patrol and Sam, their interpreter, told the Iraqi police to pull the body off our hood. They did so, laying him on the ground with looks of confusion on their faces that suggested we were idiots for going through all this trouble to recover the body of some random guy off the street. While Koky and Sam finished the conversation with the police, I provided the company headquarters with a closeout report on the situation.

When the negotiation with the police was finished and the body deposited, we continued patrolling south until we reached the abandoned seminary where my truck was parked with the rest of 2nd Platoon's vehicles.

I climbed into my truck, closed my eyes, and reflected on the situation. The weeks following our defacement of the lion were some of the busiest weeks I'd ever experienced. One of the defining features of deploying during the GWOT was the relentless pace of day-to-day life. The squadron hadn't started maintaining units in sector 24 / 7–that would come later–but we regularly had missions running outside the wire for 18 or more hours a day, every day. For the platoons, this meant they usually patrolled three to four days in a row, then took a day off to refit. For the Troop headquarters, this meant the machine never stopped running, and this was compounded by the incredible pace Captain Cook set.[2] He was a high energy leader who bounced from patrolling with the platoons, to attending squadron meetings, to conducting key leader engagements with the Iraqi police, and he had high standards for the headquarters section. In the three months I served as his XO in Iraq, I rarely saw him sleep, and I never saw him eat anything that wasn't in a Styrofoam to-go container.

The troop was still based on FOB Falcon, but all of our company's operations were conducted in *Mahala* 826, which was a dysfunctional urban wasteland. Sectarian violence between the Sunni and Shi'a populations across Iraq was a recurring problem for coalition forces. In some neighborhoods, like the Sunni dominated *mahalas* our squadron would later secure, one side clearly won and pushed the other faction out entirely. During our time in 826, though, there was no clear sectarian victor, and the entire neighborhood suffered for it. Entire swaths of the tightly packed buildings had either been reduced to rubble or vacated, and criminals and terrorists occupied many of those that remained. The few ordinary Iraqi citizens who were still living in 826 lived in fear. They avoided being outside or interacting with either us or the police, because either might result in them being killed later for perceived treachery. The Iraqi police in the neighborhood existed in a similar state of fear and alienation. They maintained checkpoints on several of the major street intersections in 826, but the police occupying the checkpoints made no real effort to secure the population, instead huddling behind reinforced concrete to avoid sniper fire or racing through the streets in their armored pickup trucks.

To fix the situation in 826, the troop needed to normalize life for the people who lived there. The government of Iraq was prepared to distribute propane, repair electric lines, and increase police presence, or at least they said they were, but to accomplish any of that, the government needed us to secure the area first. If we couldn't keep government representatives and the people they supported from being killed, they would never be able to govern the area, and the situation would never improve.

For the first few weeks, the troop tried to secure the area by patrolling from FOB Falcon, but it wasn't producing the desired effects. The area was too big, there were too

---

2. My perspective on who worked harder changed over time. When I was an XO, it seemed like the platoons had plenty of time off between missions while I was always burning the midnight oil to support them. When I was a platoon leader, it seemed like my company HQ was always screwing off while the boys and I were out risking our lives. Now, I appreciate that I was wrong in both cases. Everyone was working hard, and everyone thought he was working harder than everyone else.

many gaps in our patrol schedule, and it took too long to get a response force into sector when a problem occurred. Consequently, Crider and Cook decided to forward position the troop in the neighborhood itself. To do this, though, we needed a place to live. The place needed to be big enough to house the troop and its vehicles, close enough to the neighborhood to allow the troop to respond to problems in seconds, and secure enough that terrorists couldn't wipe out the entire compound with a VBIED. After a bit of searching, we found a place that met all the criteria, an abandoned Catholic seminary at the southern tip of Mechanics, and thus, COP Amanche was born.

COP Amanche, an amalgamation of the words Apache and Comanche—the two units that would live and operate from the base—was a beautiful building constructed in the style of a medieval European monastery with a pleasant but overgrown garden in the center surrounded by a shaded, colonnaded walkway.[3] The interior of the two-story building was lined with heavy wooden doors leading to rooms along the exterior walls. The first floor held offices, storage closets, and a chapel built into one of the corner rooms. The second floor was entirely devoted to the former occupants' rooms, and most of them still contained the beds, dressers and the odd piece of religious literature from that period. The entire structure was well maintained, if a bit neglected and in need of a good cleaning, but from a military point of view, it needed a lot of work to function as a base for the troop.

First, the outer wall had to be heavily reinforced to prevent enemy forces from firing bullets or RPG rounds through the windows or collapsing the entire structure with a VBIED. Then, the area behind the building had to be enclosed to allow vehicles and equipment to be stored somewhere they could be secured. Wiring had to be run through the largest corner office on the first floor so it could support the electronics necessary to serve as a Troop headquarters, and finally, gates had to be built to allow entry into the whole compound once it was finished. The simplest way to accomplish most of these requirements was to create a fence of Texas barriers, or T-walls as they were usually called, around the entire building. T-walls are twelve-foot tall, prefabricated, concrete barriers that look like giant versions of the concrete dividers used on the interstate.[4] FOB Falcon had a seemingly infinite supply of them, so the squadron's support company

---

3. It has since been converted into the ████ █████ ██████ (Al Hadi University College).

4. Over the course of the war, a seemingly endless number of T-walls were emplaced around Baghdad. Bases, highways, and entire neighborhoods were lined with thousands of the enormous barricades to prevent insurgent freedom of maneuver at a cost that I can't imagine.

commander, Captain Kruller, devised a plan to haul the walls from Falcon to Amanche with heavy trucks and emplace them with a crane. Once that was done, we could fill any gaps in the walls and build the entry gates with concertino wire, line all the second-story windows with sandbags, and run any necessary wiring ourselves.

The installation of the walls was labor intensive, and so far, the support company had been working twenty-four hours a day for three days straight to get the project done. While they worked, our platoons rotated through pulling security in trucks around the compound, and Comanche's infantry guys secured the roof. Captain Cook and I alternated pulling guard with the platoons and attending to the troop's requirements at Falcon. Mostly, I spent the long, slow days watching the streets and buildings in the area for signs of trouble from my truck, but to prevent enemy forces from ambushing the construction crews, our platoons had been patrolling at night to look for enemy forces on the streets of Mechanics.

I joined 2nd Platoon for one of those patrols to get a feel for the terrain around the new COP and to avoid falling asleep. Tonight, it happened to be a patrol when we found a dead body. It was four-thirty in the morning before I climbed back into the truck and recounted the tale of our patrol to my driver, King.

"That's pretty fucked up, sir. How many dead bodies is so far?"

"I don't remember. Maybe ten or twelve? Didn't 3rd Platoon find a couple last night?"

"Yeah. They found a few after that IED blew up after they walked past it. 1st Platoon found a bunch the night before too. Great neighborhood we're moving into, isn't it?"

"Yeah, this place is a fucking mess. Did D Co push any updates on how much longer they need to get the COP finished?"

"There was an update a few hours ago. It sounds like they'll be done in two more days."

Two more days. Great. It'd be good to be finished. Sitting out here all day was getting old. I pulled out my notebook and turned on my red lens flashlight and take a few notes. King spoke up again.

"Hey Sir, do you smell something weird?"

I sniffed the air. "Like what? I don't smell anything but truck and body odor."

"It smells metallic. Did you step in some shit on patrol? Whatever the smell is, I think you brought it in with you."

"I don't know, maybe?" I got out of the truck and walked around to his side. "Hit me with the white light, let's see what's up."

King opened his door and got out. His rifle had a light on it, so he pointed the rifle in my general direction and turned on the flashlight. In its blinding white light, we could both see that my pants were soaked with blood.

"Holy shit, Sir! Did you get hit?"

"No. I'm fine. It must be from that dead guy I carried. Fuck! My uniform is trashed."

It was indeed ruined. Not only had the body leaked blood all over my pants, but because I hadn't noticed soon enough, I'd put my sleeves in it when I sat down and spread it everywhere.

"What time are we rotating back this morning?"

"The commander will relieve us at 0730, sir, and I think you have a Squadron meeting at 0900."

"Yep, you're right. I need to get cleaned up. I can't go to Squadron looking like this."

I opened my door and looked at my truck seat. There wasn't too much blood on it. Seats were hard to replace, though, so I needed to keep it clean if I could. I took off my armor, stripped off my uniform top, and laid it on the seat of my truck. The shirt was relatively clean, so sitting on it would keep my pants from getting the truck seat any bloodier than it already was. I sat down, closed the door, and watched the first hint of light creep into the pre-dawn sky as I thought about what I needed to cover during the squadron meeting.

9

COP Amanche, Iraq, April, 2007

"Comanche Roof, this is Apache Five. What are you shooting at, over?"

"This is Roof. Escalation of force measures, over."

"Again? Were they demonstrating hostile intent, over?"

"They weren't slowing down properly as they approached the gate, over."

"No weapons? No nothing? Just not slowing down, over?"

There was a pause. "Roger, over."

"Goddamnit, Roof. Let me talk to your NCOIC."

The same voice replied, "This is him, over."

"Report to the TOC time now, over."

"Roger, over."

As I sat staring at the radio, all I could think of was what a shitty day it was turning out to be. It was one of those days when everything that could go wrong had gone wrong. First, our guys had smashed one of our new reconnaissance drones into a building and destroyed it while trying to get it to fly, then I'd had an argument with 2nd Platoon over how they were handling a distribution of propane to the neighborhood, *then* Muqtada Al-Sadr had said it was okay for Shi'a militias to attack Americans again, and now I was staring at the radio because I'd just used it to chew out one of the Comanche Troop NCOs pulling security on the roof.[1] The NCO was Staff Sergeant Kaluzny, one of the 2nd platoon squad leaders, and he was on his way down now to continue our discussion. I wasn't looking forward to it.

---

1. The drone was called a raven, and it was one of the early, unarmed drones the military developed to scout areas that were hard to get to. It was notoriously tricky to fly, and the one our troop smashed that day wasn't the first (or last) raven I'd had to either recover, repair, or write off.

The altercation had arisen because the Comanche guys were regularly firing warning shots at vehicles and pedestrians who got too close to the base for the infantry guys' liking. We'd had incidents with cars swerving in surprise and nasty ricochets, and I was pissed off that the guys on the roof kept taking the shots. "We're trying to get this place as much back to normal as possible," I thought. "Shooting at people isn't helping with that. It puts everyone on edge. These Comanche guys aren't getting it." In Kaluzny's defense, he and the guys were being justifiably cautious. The troop had received intelligence that insurgent forces were preparing to mass on the FOB, they had been receiving regular, sporadic fire from the neighborhood for weeks, and the local mosques had been playing a "call to arms" song all morning. The song's eerie wail was irritating, and everyone was concerned about a VBIED attack on our gate. Still, I felt that overreacting to the situation would undermine our long-term effort, and we received reports about attacks all the time, so we needed to keep our cool.

These internal tensions were a common feature of the deployment. There were plenty of guys who had no real interest in a measured response or tight rules of engagement, particularly this early in the deployment, when we didn't have much rapport with the population. "We are at war," they thought. "Fuck these people. I'm not being killed by a suicide bomber; better to kill him first." Looking back on the wars in Afghanistan and Iraq and their aftermath, I can't say the guys were wrong, but at the time, I felt strongly that we needed to be more measured and adhere to the principles of counterinsurgency.

Kaluzny walked into the operations center, or TOC, in his full battle rattle and looked at me with that special combination of professionalism and "who the fuck is this guy" that good NCOs do so well.[2] He was a huge guy with a shaved head and a build like a young Arnold Schwarzenegger, and we hadn't met before. I rubbed the exhaustion out of my eyes and started the conversation.

"C'mon, man, we can't be shooting at people like that."

"Sir, he was too close to the base. We were following the escalation of force measures in the rules of engagement."

"Yeah, I get it, but what do you think that dude you shot at is going to think about Americans? You think he's going to want to walk up to the base to tell us about any terrorists living in the neighborhood after we shot at him as he drove by? We're working

---

2. Full battle rattle is the Army term for wearing all the gear necessary to conduct an operation. In Iraq, this consisted of a weapon, armor, helmet, eye protection, gloves, knee pads, radio, medical kit, and ammunition. Set ups varied a bit by position, but it generally weighed about 50 to 60 lbs.

on winning hearts and minds here."[3]

Kaluzny's voice said, "Roger, Sir," but his eyes said, "Are we done? I've got more important things to do."

We *were* done, because this lecture would never amount to anything more. Kaluzny didn't directly work for me. He was in Comanche Troop, and I was in Apache, so my ability to influence him was limited to what general military authority dictated and what my personal relationship with him provided. As a lieutenant, I had little of the former, and with Kaluzny I had none of the latter. I knew his platoon leader, but we all had bigger things to worry about than a non-incident like this.

"Yeah, we're done. Help me out, man. We've got a war to win, and we're not going to do it by pissing people off."

"Roger, Sir." Kaluzny turned on his heel and walked back up the stairs to rejoin his squad on the roof.

I walked back across the TOC to my desk. Army doctrine says to always improve your position, and since taking over COP Amanche, we'd done so. The TOC now had a full bank of phones, radios, and computers sitting on plastic folding tables along one wall. Over the bank of communication equipment was an air conditioning unit wedged into an old window with plywood and sandbags to maintain a livable temperature in the room. Nobody was too concerned with the troop's comfort, but the electronics would shut down if the room got too hot, and that wasn't acceptable. The rest of the room was a hodge-podge of old couches, equipment boxes, and weapon racks. The walls were decorated with posters depicting scenes from *Starship Troopers* with retooled slogans created by the Comanche Troop operations sergeant, Bill Highsmith. A professional and capable guy with a knack for digital art, Highsmith had dubbed our base "Fort Joe Smith" after the base in *Starship Troopers* that was overrun by arachnids.

As I sat down, Highsmith turned to me and said, "Kaluzny is a solid NCO. He gets it. The guys are just frustrated with the situation."

"Yeah, I understand their position. It's fine. I'm just pissed off today."

Captain Cook, First Sergeant Strong, and the Comanche command team were back

---

3. According to Wikipedia, 'winning hearts and minds' is a concept occasionally expressed in the resolution of war, insurgency, and other conflicts, in which one side seeks to prevail not by using superior force, but by making emotional or intellectual appeals to sway supporters of the other side. Over the course of the GWOT, the phrase became first a cliché, and then later the tongue-in-cheek punchline to an endless series of jokes. Of course, even after being reduced to a joke, it remained official policy.

at Falcon, so I was running the show today. The propane distribution 2nd platoon had run this morning had been a mess. Propane distribution was an essential service to the people in Baghdad. The gas was the primary means by which most people cooked food and heated water. The old regime had regulated the distribution of propane through a system of ration cards. Every family had a card, and by presenting that card to the propane distributors as they drove through the *mahala*s, a family could pick up its allotted can. When the regime collapsed, that system broke down. The offices that issued ration cards became unreliable, so when people used up or lost their cards, their ability to get replacements was limited. Additionally, as sectarian violence exploded across the country, propane distributors halted their trips through neighborhoods to distribute cans directly to people, so those without cars, donkeys, or some other way to get propane cans from the distribution centers to their homes, couldn't get propane at all.

When Apache Troop took over in Mechanics, one of the missions the platoons regularly undertook was to provide security for propane distributors as they made their deliveries. Crider and Cook correctly assessed that this was an effective way to build goodwill with the people, but the missions presented problems we hadn't foreseen, like the one 2nd platoon encountered during this morning's operation.

2nd Platoon's plan had been to start the distribution mission early, driving to the propane bottling plant just north of FOB Falcon to pick up the distribution teams, each of which had a tractor that pulled a trailer full of propane cans. From there, the platoon would escort the tractors into Mechanics, set up a basic perimeter of concertina wire, and then manage and secure an orderly distribution process. The reality turned out to be a bit different.

Because the population of Mechanics had been without propane for so long, as the convoy moved through the neighborhood, it attracted a huge following of people on foot. Because the convoy was slow, by the time it arrived at the designated site, there was an enormous crowd of people there. This made establishing a perimeter difficult and slow, as the platoon had to physically push people back to lay the wire and set up a distribution line. This process caused numerous small arguments as people squabbling for position were jostled against each other or yelled at by the interpreters. It was also hot outside, and by the time the distribution finally started, the crowd was already agitated and irritable.

The mission hit its next snag when an old woman arrived at the front of the line without a ration card. The distributor refused to provide her with propane, and she began to wail and sob. She'd been unable to cook for days, and now it looked like she'd be going home to wait longer. 2nd platoon escorted her to the exit, but she didn't leave, and as more people without cards were turned away, they grew into a yelling, furious mob, particularly as the people without cards who were still in line realized they wouldn't get propane that day either and joined in to express their anger.

Trying to diffuse the situation, 2nd platoon negotiated with the distributor to provide propane to people without cards, but the distributor insisted he wasn't authorized to do so and that he'd get in trouble if he did.[4] 2nd Platoon called the troop headquarters to request guidance on the situation before they shut down the distribution for the day. Everyone with a card had received propane, but most of the people who had come didn't have cards and hadn't received any, and there was still plenty left on the trailers.

I picked up the hand mike when 2nd Platoon called, and SFC Edgy explained the situation.

"Apache five this is White-four. They aren't going to distribute any more propane today. None of the people here have ration cards."

"White-four this is five, I'm tracking that, but aren't there still a bunch of people who didn't get any propane?"

"Yes, sir, and the crowd is pretty pissed off about it. Still, the distributor won't back down. I recommend we terminate the distro and take this up with the distributors' leadership at the bottling plant."

I could hear the crowd in the background on the radio as Edgy spoke. I put the hand mike down for a second and thought. What was the right answer here? On the one hand, people needed propane, and they were unlikely to support either us or the Iraqi government if we couldn't provide essential services to them. On the other hand, the distributor *was* the Iraqi government, and if we undermined the existing system, we risked the distributors refusing to cooperate with us in the future, further weakening an already shaky process.

One of the most challenging parts of the war was trying to figure out how to solve these sorts of problems. The oldest guy on COP Amanche that day was probably me, at twenty-six. I had a degree in philosophy, six years of experience as an infantryman in and out of combat, and zero knowledge about how an American city functioned, much less

---

4. A year and a half-dozen meetings with the heads of the regional utility managers later, I never got a straight answer on exactly why this was a problem. I assume it was profit driven.

an Iraqi one. The Department of State had people in theater who were experts in this sort of thing, and the Squadron even had a civil affairs captain who could have probably provided a solid answer. But there were never enough of these sorts of specialists to go around, so the guys on the ground rarely had access to their expertise when they needed it. There was a mob of people on the ground who might get violent if they didn't get propane, and 2nd platoon needed my best answer now, not a perfect one later.

"Alright White-Four, this is five. Have your guys distribute the propane themselves. Fuck the distributors."

There was a pause as SFC Edgy processed this. Edgy was a smart guy, and he was doing the same mental math I just did. Because he was the guy on the ground, his situation was even more difficult. Whatever impact the decision made would be felt by him and his guys then and there, and that was at least going to be a miserable social encounter, and it might escalate into something worse.

"Apache five, is Apache six there?"[5]

"Nope. six is back on the FOB. I'm it today."

"Apache five, are you ordering me to distribute this propane over the objections of the Iraqis?"

"White-Four this is five. Yeah, I'm ordering you to distribute the propane."

"Roger that. White-Four out."

Edgy sounded pissed, and he had good reason to be. He was about to spend an hour fighting against the dysfunctional Iraqi utility distribution system, and in the process, his patrol timeline was probably going to get screwed up, which would mess up his entire reset schedule and cut into his guys' chow time. The next call I got was an hour later. All the propane was distributed. The distributors were furious. 2nd platoon was escorting them back to the bottling plant, then the platoon was coming back to the COP to reset and prepare for their next mission that day: escorting the D Co wreckers that were continuing to emplace T-Walls around the major streets in the neighborhood to keep insurgents from burying IEDs under them.

Irritated with the situation, I set down the hand mike and walked to the chow tent for lunch. As I waited in line, wondering whether I'd made the right call on the propane problem, the Comanche troop guys on the roof took the aforementioned warning shots.

---

5. Apache six was Captain Cook's callsign. Apache five was mine. Red Platoon was 1st Platoon, white platoon was 2nd Platoon, and Blue Platoon was 3rd Platoon.

By the time that argument was resolved, the chow tent was shut down until dinner, and the only remaining lunch option was a heater meal.[6] I grabbed a can of Cajun beans and rice and sat back down in the TOC to eat it with tabasco sauce, an Otis Spunkmeyer muffin, and a bottle of water. COP Amanche had plenty of pallets of Cajun beans and rice, but not too many other ready-to-eat options. I assume other flavors of the heater meals existed, but I can't confirm that, because we never saw them.

To this day, I suspect that somewhere in Iraq there were entire shipping containers of delicious food, but fobbits above us in the logistics chain always managed to siphon them off, so all that was left by the time the supplies got down to our level were cans of Cajun beans and rice.[7]

I took my can of food into the courtyard, opened it, and chewed on both the situation and lunch. Counterinsurgency was a pain in the ass. How were we supposed to provide security when the population hated us and our Iraqi partners so much that they supported the insurgents who attacked us? But then again, how were we supposed to get the population to like us without providing the security necessary to get them the services they needed? I puzzled over this problem as I finished lunch, then walked back to the TOC to work through our weekly supply requests. Our supply guy was coming out later that day with the D Co logistics package, and he needed to know what they needed to bring.

That math, and the accompanying walk-around trips Highsmith and I took to count our existing supplies, took a few hours, and I was just inputting the last few numbers on the logistics spreadsheet we maintained when the radio speaker erupted in sound. It was Lieutenant Sisoura, and he sounded stressed out.

"Apache X-Ray this is White-One. Incoming MEDEVAC request. Prepare to copy ."[8]

The group of us standing in the TOC stared at the radio blankly for about half a second, then there was a frenzy of movement as Highsmith, Bonilla, and the rest of the

---

6. Heater meals were canned meals with a built-in heater. Picture a can of soup glued to the top of a can of Sterno.

7. Fobbits, or their predecessors in other wars, the REMFs and POGs, are the boogey-men blamed by the guys on the front lines of our nation's wars for taking all the good stuff that was meant for the *real* soldiers. Fobbits were also scorned for living in relative comfort and safety, and generally milking their wartime service for all the enjoyment and profit they could. The perspective had a kernel of truth, but most of the front-line folks wouldn't have had it any other way and bitching about the disparity was just part of the fun.

8. X-Ray is the standard callsign we used for our HQ.

crew scrambled for paper and pens. I grabbed the hand mike and keyed it.

"White-one this is five, send it."

Koky's voice had leveled out, and he calmly pushed the nine lines of the Army's standard medical evacuation format. The nine-line report communicates, as concisely as possible, the number and condition of casualties received, which medical evacuation (MEDEVAC) assets are needed to transport them, and where the MEDEVAC assets should come for the extraction. In this case, 2nd platoon had suffered one seriously injured casualty and needed a helicopter to land in the field next to COP Amanche to recover him immediately.

The thirty minutes in the TOC following Koky's radio call were pandemonium. The second I put down the hand mike every phone in the TOC rang. I picked up one of them and balanced the receiver on my shoulder as I gestured to Highsmith and Bonilla to answer the others.

"This is Lieutenant Pace, what do you need?"

"Hey Lieutenant Pace, this is Bolson from the Squadron TOC. I just heard the radio call for a MEDEVAC, and I was wondering if you could give me a few details on it."

Bolson's voice was painfully slow and deliberate. As he spoke, the radio crackled to life again. Bonilla and Highsmith were already on the other phones, so I spun around the room looking for someone else to answer the radio. I saw Banninger and pointed at him to grab the mike. He did so and started talking.

"This is Apache X-Ray. Go ahead over."

Bolson's voice continued. "The commander is on his way up here now, and I want to make sure I can spin him up on the situation when he arrives."

"Okay, Bolson, I'll get you the information, but I'm a bit busy here at the moment. Let me work through the details first."

Bolson's response was long, and in it he emphasized the need for fast, accurate reporting. His drawl was difficult to hear over the increasingly loud conversations around me as everyone's volume escalated so they could be heard over the developing cacophony. Their voices rang in my ears as my brain tried to place each one with its respective speaker and problem.

Banninger's voice. "Roger White-One, I'll grab Apache five."

Bonilla's voice. "Hey Sir, the MEDEVAC TOC is on the phone. They'd like some details on the situation."

Highsmith's voice. "Sir, Captain Kruller is on the phone. She needs an update on her team that's with 2nd Platoon."

"Hey Bolson, I gotta go. I'll call later."

"Banninger, give me the hand mike."

"Bonilla, give them the 9-line info and tell them we'll push updates when we have them."

"Highsmith, tell her I'll call when I can. I'm sure they're tied in w/ 2nd platoon."

"White-One, this is Apache five. Go ahead."

"BREAK BREAK, Apache X-ray, this is Comanche two-six, contact IED over."

One of the phones started ringing again.

"Baninger, grab that." Then over the radio, "Two-six, this is Apache five, go ahead over."

The other phone rang. Highsmith grabbed it.

"Apache five, this is Comanche two-six. White-Four hit an IED. Their truck is a mess, and they have casualties. We're developing the situation now."

Highsmith's voice. "Sir, the Squadron wants to know if 2nd Platoon is talking about a second IED or still the first one."

Banninger's voice. "Sir, the Squadron medics want to know how many casualties are incoming."

I waved my hand at them both irritably. "Tell them we're working it." Then over the radio, "two-six this is X-ray, do you need any support?"

"X-ray, this is two-six, we're going to need..." I couldn't hear the rest of the call over an explosion of radio static from the other radio and the second phone ringing again.

"Bonilla! Turn the volume down on the radio and answer it. Banninger, grab the other radio." I keyed the hand mike I was holding. "Two-six, this is X-ray, say again over."

Two-six sounded irritated now. Nobody enjoyed repeating himself. "X-ray, I say again, we're going to need a wrecker to recover this truck, and we need the QRF spun up to secure it–I don't have any more available trucks."

"This is X-ray, roger, I'll work it."

"Apache five, this is Comanche two-six, MEDEVAC request to follow. Prepare to copy."

"Comanche two-six, send it."

"Apache five, updated casualty count to previous MEDEVAC request. Add two litter urgent and one possible KIA."

"Roger. Updating MEDEVAC request. You got the location?"

"Roger."

I turned to Highsmith. "Call the Company TOC on Falcon and get 1st platoon spun up on the situation. They need to grab a wrecker and get to 2nd Platoon's location." To Bonilla, "Update the MEDEVAC bird with three more casualties. Same pick up location."

Bonilla picked up the radio to talk to the birds, and Highsmith nodded, but as he moved to pick up the phone and call Falcon, the phone rang again. He answered.

"Sir, it's Captain Cook."

Bonilla's voice. "Sir, the MEDEVAC bird is on station. ETA, ten minutes."

"Highsmith, relay that to both platoons. I gotta take the boss's call."

I swapped my radio hand mike for Highsmith's telephone, tangling both cords, and spoke into the receiver as we danced around one another to untangle ourselves.

"Sir, this is Dan."

"XO! I heard what's happening. I'm spinning up trucks from Falcon now, what do you need out there?"

"I need 1st Platoon to pick up a D Co wrecker and link up with 2nd platoon. They've got two downed vehicles and at least four casualties, and the platoon is split. Half in the north, half in the south."

"Where are the casualties?"

"Three in the south with Edgy and one in the north with Sisoura. The wrecker needs to get to Sisoura in the north though."

"Got it. I'll work it. I'll call once we're up on the radio."

"Okay, Sir."

I hung up the phone. It rang again immediately. Ignoring it, I turned to Highsmith. "What's up with 2nd platoon?"

"They're working on setting up the LZ for the bird now. They're setting it up right outside the COP in that field east of us."[9]

"Do they need help?"

"Probably."

I thought for a second and said, "Take a team of guys and walk out the gate to meet them. Get with Comanche and see if they can spin up a few more trucks from here

---

9. An LZ is a landing zone, which is a space large enough and sufficiently clear of obstacles to land a helicopter.

to help secure the LZ." Then to the room in general, "Somebody grab that goddamn phone!"

Highsmith nodded and ran out of the room. Banninger grabbed the phone. Bonilla was still on the other one, receiver in one hand and radio hand mike in the other.

I keyed the hand mike in my hand and called 2nd Platoon. "White-one this is Apache five, can I get an update? You guys need anything?"

The voice that came back was unfamiliar. Sisoura was probably busy, so one of the drivers responded.

"Roger Apache five, this is White-one-delta. White-one actual is busy working on the truck, and one of the guys got banged up pretty bad from the blast."

"Do you need MEDEVAC assets?"

"Wait one."

Bonilla's voice. "Sir, Bolson wants an update on the situation, and the MEDEVAC guys need an update on the LZ security posture."

The radio. "Apache five, we may need a second MEDEVAC, I'm waiting on White-one actual to give me guidance now."

All three voices came through in a simultaneous snarl. My irritation flared as my brain worked to process it sequentially and make sense of it.

"Tell the MEDEVAC crew we'll have the LZ secured by the time they're here, and tell Bolson I'll call his ass when I know more."

Bonilla nodded and started relaying information over the radio.

The phone rang again. Banninger answered it.

The radio keyed in my ear. "Apache five, this is Comanche two-six. We're set at the LZ and ready to receive the MEDEVAC bird."

Banninger's voice. "Sir, Captain Kruller wants a status on her guys."

For some reason this was the straw that broke the camel's back. I turned to Banninger and lashed out.

"Tell her I'll fucking call when I figure out what their status is! I'm sure they're fucking fine! We have people dying here, and we need to work through that shit before we figure out the rest of what the fuck is going on with all the other shit!"

Banninger's voice. "The phone went dead, sir. She hung up."

I keyed the hand mike. "Roger two-six, we've passed the MEDEVAC bird to your net. You should have comms with them now. We've got guys from the COP reinforcing your guys at the LZ, and Apache six is *en route* with the wrecker to your element up north."

"This is two-six, roger."

The phones continued to ring. I set down the hand mike and took a moment to compose myself. I knew shouting wouldn't help the situation and that I needed to calm down a bit.

As with so many other aspects of our deployment, the MEDEVAC process eventually got smoother, but this was Apache Troop's first, and it was a mess. As we stumbled through the fog of war and fought to understand the situation, we eventually learned that while driving north through Mechanics to meet the D Co T-Wall emplacement team, one of 2nd platoon's trucks had hit an IED buried under the street. The bomb had damaged the truck severely and injured one of the guys riding in it. While responding to the blast, another 2nd platoon truck–Edgy's–had hit a second IED. The bomb had detonated under the driver's seat of the truck, propelling Specialist North, the truck's driver, into the armored roof of the truck. The blast left him dead and two others injured, including the platoon medic. Only Edgy had walked away, dazed but functional. Fortunately, the enemy hadn't conducted an ambush in conjunction with the IED strikes, so with the help of Comanche Troop, 2nd platoon had been able to secure both damaged vehicles and coordinate the MEDEVAC. They set up the helicopter landing zone in the field east of COP Amanche, secured the area with their vehicles, and talked the MEDEVAC bird onto the site. Meanwhile, the Comanche Troop medic prepared the casualties for transportation, providing onsite treatment, lashing them to litters, and stabilizing them for the flight. When the helicopter arrived, a mixed party of 2nd platoon and Comanche Troop guys ran the litter out to the helicopter and loaded the dead and injured on board. The MEDEVAC went smoothly, and the casualties were flown to the Combat Surgical Hospital, or CSH, in the Green Zone. All the other Apache soldiers recovered, but North was beyond help. Specialist North, the great kid we'd trained with, joked with, and fought with for the last year, was the first Apache Troop soldier killed in Iraq.

Captain Cook and First Sergeant Strong drove out from Falcon to take charge of the situation, and while they managed the operation to get the vehicle recovered and to have 1st platoon replace 2nd platoon in sector, things started to slow back down. I sat down, found our book of policies and procedures, and started working through the admin process that accompanies a vehicle being destroyed and a soldier being killed.

The tone of the COP the rest of that day was somber. The headquarters guys who had managed the situation were all going through the motions of their duties in the sluggish, heavy-lidded, semi-stupor that follows a serious adrenaline rush, and there wasn't much talking in the TOC. Fortunately, there wasn't any more excitement as we worked through the necessary processes, answered the questions that came over the phone, and tracked 1st platoon's periodic radio calls. The evening continued in the same manner, and around dinner time, Cook, Strong, and 2nd platoon returned to Falcon to reset. A few hours later, when the night shift took over and the last patrol was in for the night, I sat down on the couch in the TOC to talk to Highsmith.

"Well, man, that was a hell of a day."

Highsmith nodded. "Yes, sir. It was. It's a damned shame about North."

"Yes, it is."

We sat in silence for a while, because there wasn't really anything more to say about it. It was a damned shame. North had been a great guy. He had been optimistic, capable, and hard-working, and now he was dead. Meanwhile, it seemed like we were no closer to securing Mechanics than we had been the day we arrived.

"Oh, by the way, sir, while you were eating, the call came down from Squadron that our deployment has officially been extended from twelve to fifteen months."

I nodded silently in response to the news. It had turned out to be an even shittier day than I thought.

10

FOB Falcon, Iraq, May, 2007

Five-inch holes riddled the armored Humvee's bulletproof windows and the steel shield that surrounded the turret to protect the truck's gunner, and everything inside the vehicle was coated in slag, powder burns, and blood.

"Alright, Miller, talk us through the process of cleaning this thing out and getting the equipment accounted for."

Sergeant Miller, Specialist Maine, and I were standing in the troop motor pool on FOB Falcon. A 2nd Platoon Humvee was sitting in front of us with its trunk open. Like all the trucks we used in Iraq, the Humvee was heavily armored, but this truck's armor had been defeated by a new type of IED, the explosively formed penetrator, or EFP. As far as IEDs went, EFPs were high tech. They were an old technology that had long been used in anti-tank weapons but had been adapted recently for use in IEDs by the wealthy sponsors of the Shi'a insurgents in Iraq. When an EFP detonated, it fired a super-heated copper projectile that could cut through an armored Humvee like a hot knife through butter. Moreover, because EFPs were often set off by a passive infrared system, could be hidden easily in the huge amounts of trash along the roads we patrolled, and had a range of up to 50 or 60 yards, they were almost impossible to detect during regular patrols.

2nd Platoon had hit this EFP yesterday. It had cut through the truck, punched a hole through the gunner, Specialist Dixon, and continued through the armor on the other side of the truck without pause. The platoon had cross-loaded Dixon into another truck and rushed him to the CSH, and initially their effort looked promising. When they unloaded Dixon at the hospital, he'd been conscious, and the platoon had been optimistic about his recovery since they'd made it within the "Golden Hour," a planning guideline that said if you could get someone to a major hospital within an hour of being injured, his chances of survival were significantly higher. Unfortunately, the EFP had done too much damage to his internal organs, and he'd died in surgery.

The damaged truck had to be repaired, and the gear inside had to be inventoried, but before any of that could happen it had to be cleaned. Normally, that process was platoon business since the platoon leader owned the property, but I found Koky and told him we'd take care of this one. Dixon was a well-liked member of 2nd Platoon, and he'd been killed not long after North, so the platoon had taken his death badly. It didn't seem right to me that they should have to clean his remains out of the truck, so Miller, Maine, and I took the inventory forms and did it for them.

"I appreciate you helping me with this, guys. I know it's not the most pleasant job, but we can't have 2nd Platoon's guys doing it."

Miller was a quiet, professional supply sergeant and a devoted family man. Maine was an intense, but hard-working and creative cavalry guy who we'd pulled into the supply room to help run it. They had both known Dixon well, and they agreed it would be wrong to make the platoon take care of this.

Cleaning the truck was a grisly process. The EFP had filled the cab with hot slag, powder burns, and blood, and everything inside the vehicle was coated in all three. We started by pulling out the equipment that had been ruined by the blast or was so covered in blood we couldn't use it anymore, inventorying it, and throwing it away. Next, we catalogued the equipment that wasn't ruined and set it aside for further cleaning. Finally, we drove the truck to the wash racks, sprayed it out with a pressure washer, and turned it into the motor pool to be repaired. The motor pool sergeant processed the paperwork to remove the truck from our books, and he gave us a new vehicle so the platoon could continue patrolling while the old one was being repaired. The whole process took about two hours.

I looked at Maine and Miller when we were finished. We were covered in filth and soaked with water, sweat, and blood. Dixon's vehicle was scrubbed clean, but the process had left us dirty, tired, and emotionally distraught. Cleaning the truck had been the right thing to do, but it hadn't been enjoyable.

"We did a good thing here, guys. Get cleaned up and take the rest of the day off. Dixon's memorial is this afternoon at squadron."

They nodded, quiet and on the edge of tears. We walked back to the troop, picked up fresh uniforms from the supply room, and went our separate ways to get cleaned up.

The memorial was a somber affair. It was held in the FOB Falcon chapel. The chapel was a small, two-room building with a plywood door and two rows of pews facing an altar that had been adorned with a photo of Dixon on a small table in front of his boots, rifle, helmet, and dog tags, which were arranged in the traditional military way. Our chaplain led the squadron in a prayer, Dixon's friends got up to say a few words about him, and a bugler played taps as a firing squad fired a twenty-one-gun salute. The ceremony concluded with the singing of *Amazing Grace*, and afterward, we all filed out past Dixon's photo to pay our quiet respects. Dixon had always talked to me about going to Ranger School, so I took off my tab and left it in his upturned hat on my way past the display.[1] I left the memorial with tears in my eyes and walked back to the barracks alone, saddened that this was as close to Ranger School as Dixon would ever get.

The war didn't stop for casualties, though, and by the time I was back at the barracks, my grief was buried under the weight of everything else that had to happen today. With the memorial over and the truck reset, I needed to get back to COP Amanche. Captain Cook had to stay at Falcon this afternoon for a squadron meeting, and we preferred to have at least one of us at the COP to deal with emergencies and maintain command presence there. Normally, I'd drive out with one of the headquarters guys on a resupply mission, but 1st Platoon was about to leave on a patrol, and they were heading to Amanche afterward, so I decided to tag along with them on the mission. I grabbed my bag, found Lieutenant Humphrey, and walked down to the motor pool with him for the patrol brief.

"Where are we headed today, Rob?"

"We're going to meet with one of the Iraqi police generals to talk through improving security in Mechanics, then we're going to push a dismounted patrol up through the ruins on the western end of the neighborhood."

These ruins were a kilometer or so of old, one-story, roofless stone buildings that were arranged in a neighborhood or sorts that ran north to south along the western border of Mechanics. The platoons had taken fire from the ruins several times, and we suspected caches of weapons or IED material were hidden in the area.

"SFC Murray is taking charge of the dismounted patrol, so I'd like you to walk with him and maintain comms with the troop. I'll parallel you guys with the trucks on the

---

1. Because of its role in memorial ceremonies, I now have a viscerally negative response to hearing *Amazing Grace*.

road east of the ruins and help you out if you get into anything."

I nodded, and we finished the walk to his trucks. When we arrived, Rob's guys had everything loaded and were ready for the patrol brief. Rob talked to Murray and Curtwright for a few minutes, then stood on the hood of a truck and read the mission for the day, periodically calling on his guys to provide key pieces of information to ensure they knew the material as well as they should. The brief took about five minutes, then we loaded the trucks and rolled out. I was in Rob's truck, and his driver and gunner joked and kidded with one another over the truck's intercom system as we drove the mile or so to the FOB's exit.

When we reached the gate, a serpentine constructed of HESCO barriers and concertina wire which was flanked by enormous, twenty-foot sandbag towers, we paused to lock and load our weapons, and everyone's demeanor changed.[2] To get to Mechanics from Falcon, we had to drive up Route Jackson, the same road where 2nd Platoon had hit the EFP that killed Dixon. Despite the route clearance teams' best efforts, Jackson remained a dangerous road, even right outside FOB Falcon, and units regularly hit IEDs there. The shoulders of Jackson were kept clear of garbage, but fields full of trash paralleled the road, and that trash often hid explosives. There weren't a lot of alternatives to taking Jackson though, and those that existed were also dangerous and usually slower, so our platoons generally just tossed the dice and hoped it wasn't their day to roll snake eyes. Today, 1st Platoon was lucky, and we made it into sector with no issues.

We wound our way through Mechanics until we arrived at the small traffic circle where we were meeting the general. Rob placed his four trucks in a line, arranged his security guys, and waited for the Iraqi commander to arrive. We didn't have to wait long. The general was in sector today on one of his rare patrols, and when he arrived for the meeting, he arrived in the center of a convoy of fourteen machine-gun armed, armored pickup trucks. He arranged his trucks in an L shape, with his vehicle at the vertex near where our trucks were parked, and stepped out of his vehicle. Lieutenant Humphrey, SFC Murray and I got out of our trucks to meet him.

I placed the general's age at about fifty. He was tall and heavyset, with a camouflage uniform and cap, and a pistol on his belt. He was flanked by two Iraqi policemen wearing armor and carrying rifles, and his demeanor was agitated. The general seemed (rightfully)

---

2. The HESCO barrier is a collapsible metal frame lined with woven cloth. When unfolded and filled with dirt, HESCOs form an explosion and impact-resistant brick that can be used to build a wall. HESCO walls are used to build FOBs around the world.

concerned about his safety in this neighborhood, and his agitation was manifesting itself through large hand gestures and a loud, aggressive tone of voice.

He and Rob started talking, with Rob's interpreter translating back and forth between them. Rob asked him questions about the Iraqi police patrol schedules, the number of operations they'd conducted recently, and the number of arrests they'd made. The general provided vague answers and continued to speak in a loud, aggressive tone that suggested he didn't think much of us or our questions. As the conversation continued in this tone, I could see that SFC Murray was getting increasingly agitated. He didn't have a calm temper under the best of circumstances, and this general was clearly getting under his skin. A few minutes later, when the general responded to one of Rob's questions by flipping his hand in the air and turning his back to Rob, SFC Murray had clearly had enough. He stepped forward and entered the conversation.

"Excuse me, sir, but I need to talk to this mother fucker for a second." SFC Murray continued forward, walking past Rob and standing directly in front of the Iraqi General. Murray was about six inches taller than the general and muscular whereas the general was fat. He placed his face about two inches from the general's face and started shouting at him.

"You better respect my Lieutenant, you sorry mother fucker! This ain't my country we're over here defending, it's yours. And you gotta lot of nerve disrespecting my commander when we're over here sweating and dying for your sorry ass!"

At this point, SFC Murray was pretty wound up, and he continued his diatribe while the Iraqi general visibly quailed under his wrath. SFC Murray's position, while not presented very tactfully, reflected the perspective of many of the guys, including me. The Iraqi police never seemed nearly as interested in risking their lives to defend their country as we were, and it infuriated most of us at one point or another. As I came to know the individual police officers better, we discussed this issue directly over lunch many times, and in the Iraqis' defense, for sectarian reasons, many of them didn't view securing these neighborhoods as their problem. The average Iraqi police officer was a poorly paid Shi'a who hated, and was hated by, the Sunni residents of Mechanics. The Sunnis had supported the old regime, and under that regime they had murderously oppressed the Shi'a population of Iraq for decades. As the new government was mostly Shi'a, in many Iraqis' view, the tables were now turned, and it was the Shi'a population's turn to be on top. Moreover, patrolling Sunni neighborhoods was a dangerous prospect for the police. Compared to their American counterparts, the Iraqi police had worse equipment, worse logistical support, and negligible medical resources. Consequently, they fared much worse than we did when they hit IEDs or were ambushed by insurgents. None of this made the disparity in effort between our elements any easier for most American soldiers to swallow, but in hindsight, the disparity did have a rational explanation.

The shouting continued for a minute or so, which, given the circumstances, seemed uncomfortably long. I didn't think SFC Murray would take a swing at the general, but I couldn't entirely rule it out, either. As I stood and considered this, I looked around to take in the situation. The general had fourteen, machine gun-armed trucks full of police. That worked out to about sixty guys. We had four trucks and about twenty. The Iraqis were in a better position, although our guys were more aggressive and better trained. If this escalated too much, I wasn't entirely sure we'd come out on top, and from a mission completion point of view, it wasn't a fight we wanted to have, regardless. I pointed this out to Rob, and he nodded, working through the same math problem I had. Armed with this knowledge, he stepped up, put a hand on SFC Murray's shoulder, and told him he should probably back off for a minute. Murray paused and looked at his lieutenant. They exchanged a few seconds of nonverbal communication, and Murray nodded, walking back over to the trucks to cool off. Rob and the interpreter wrapped up the meeting with an agreement that the police would patrol more often, and the Americans would check in with them before driving into sector. After a few strained parting pleasantries, Rob and I shook hands with him, and he walked back to his truck and drove away.

"Well, I'd say that was rather less cordial than he was used to."[3]

Rob laughed. "Yeah, SFC Murray was pretty wound up. We'll probably hear about that from squadron when he goes home to tell his American counterpart what happened."

I shrugged. "Yeah, probably so, but fuck him. Murray was right. Maybe it'll make the Iraqis patrol a bit more."

Rob nodded. "Maybe so. Probably not, though."

"Ha. Yeah, probably not."

We walked over to talk to SFC Murray, who was standing next to Curtwright's truck, laughing with his section leader, who was recounting the meeting from his point of view.

---

3. *Braveheart*, 1995. My head is crammed with old movie quotes that spring forth at all the worst times. If I could replace them with more useful information, I'd probably be a rocket scientist instead of a retired ground-pounder.

As we walked over, they turned to Rob, and Murray addressed him, shaking his head.

"Sorry, sir, I got a little fired up back there. That shit pissed me off, though."

"No apology necessary. I get it. Is the platoon ready for the patrol?"

"Yes, sir, the boys are ready to go when you are."

The boys were ready. SFC Murray, Johnson, and Irizary, were fully outfitted with the weapons and communications they needed for the dismounted sweep north through the Mechanics ruins. Every man had on his IBA (Interceptor, Multi-Threat, Body Armor–33.1 lbs.), ACH (advance combat helmet–2.5 lbs.), rifle (9 lbs.), 210 rounds of ammunition (7 lbs.), Harris 152 radio (2.7 lbs.), IFAK (Individual First Aid Kit–1 lb.), and a backpack with his share of the squad's medical gear, breaching tools, grenades, and emergency marking panels (10-15ish lbs.). Some guys carried water. Despite the infamous *Blackhawk Down* incident where an entire platoon found itself stranded with nothing to drink, most guys still didn't bother. Sixty-five pounds was already a heavy load to run around with, and the trucks wouldn't be far away.

We had a final talk-through with Rob on the details of the plan and set off west toward the southern end of the ruins. Murray's guys had the lead. Murray walked immediately behind them to control the patrol, and I walked behind him to maintain communication with the troop. After making it about 400 meters, we found the short wall that marked the edge of Mechanics, climbed over it, and moved into the dusty ruins to start our movement north.

The weather that day was hot and still, as always, but not as dry as you might think. Mechanics was located just to the west of the Tigris River, and that kept enough humidity in the air to make it steamy. Dismounted daytime patrols in Iraq were always a race against dehydration. Moving with sixty-five pounds of gear in one hundred ten-degree heat meant guys were sweating heavily before they even started the movement, and by the end of any foot patrol, everyone's uniform was always soaked through.

Our movement was slow and deliberate. The ruins were single-story stone structures with plenty of windows and doorways, but few actual doors–not entirely unlike the replica buildings in the Army training areas where the troop had worked through its urban movement techniques. We cleared the buildings as we came to them, finding nothing in the first few beyond broken glass and debris that suggested people had been camping in them. Clearly these buildings had been abandoned for a long time.

We continued north, reporting our position to sync our movement with Rob's mounted element, and continuing to check every crevice, alley, and hole we came across.

We'd just moved half the patrol across a small road that ran between two buildings to clear a room, when the walls south of us suddenly erupted in splashes of dust and the air filled with the whine and crack of small arms fire. Somebody from the north was shooting at us. We all spread out and took cover against the wall on the north side of the road. SFC Murray had pushed across with the first half of the patrol, so we were now facing each other on opposite sides of the small road. We peeked around the corner to see where the fire was coming from, taking several more rounds in the process. I couldn't see the shooter, but looking north, there were a dozen or more buildings' worth of darkened doors and windows. The shots could have been coming from any of them. The movement north would put us directly in their line of fire; we'd need either concealment or suppression to advance and clear the bad guys out of their position. SFC Murray had already directed Johnson to clear a small shed behind us to ensure we wouldn't get caught in a crossfire. I turned to look for the mounted element. There was one truck I could see, and it was parked behind the small wall we'd climbed over, unable to get to us for the moment. Having no platoon radio, I waved and yelled to its gunner.

"Get your ass over here! We need suppressing fire!"

The gunner, Specialist Phillipus, looked down and relayed the message to his truck commander before replying. "Sir, we'll fuck up the truck driving over this wall!"

"I don't give a shit about truck damage, I'm the XO! I'll buy you a new one!"

He nodded, and after a few seconds, the truck's engine roared, and it broke through the small stone wall, rolling up to our south and opening fire with the M240B machine gun on its turret. I looked at SFC Murray, who was smiling at me and shaking his head.

"Hell of a day, sir."

"Hell yes, it is. You got smoke?"[4]

Murray nodded, and talked to Johnson, who fired two smoke rounds up the road from his grenade launcher. Once the smoke built, we'd be able to push forward with a bit of concealment. In the seconds it took for the smoke to billow into a full cloud, the truck stopped firing. I looked back, and the gunner was nowhere to be seen.

"Hey Flip, what are you doing!?"

Phillipus popped his head up. "Sir, I just got shot in the head!"

I thought for a second. He didn't look like he was bleeding. "Well, are you alright?"

---

4. The military uses white smoke grenades to create clouds of concealing haze that prevent enemy forces from seeing through it. Colored smoke grenades are instead used to signal various things in situations where radio communications can't describe them clearly, such as marking a specific spot for a helicopter to land.

"Yeah. I think so."

"Then get your ass up there and start shooting!"

Phillipus grinned and shouted, "Yes, sir." He started firing again.

The red and white smoke had built up by now, and I looked across the road at SFC Murray.

"You ready to move, sir?"

I nodded. "Yeah, Rob in position too?"

He nodded, and we ran together up the alley as the truck fired over our heads. As we ran north, the firing toward us stopped. The guy must have changed positions. We heard Rob's trucks east of us firing into the ruins ahead of us; he was using the smoke we'd thrown to mark our forward position. We cleared through the next set of buildings, finding spent shell casings but nothing else, and paused briefly to assess the situation. SFC Murray was on the radio with Rob. Phillipus's truck was still at the southern end of the small road; the truck was too big to follow us further. The rest of our dismount squad was peering up and down the road from our building to watch for movement. With his coordination call complete, SFC Murray turned to me.

"Sir, the LT is continuing his movement north. We've still got a lot of ruins to go, and that asshole could be anywhere."

I nodded in agreement, and we cleared another dozen buildings over the course of about 200 meters. We didn't take any more fire. Clearly, whoever was shooting at us had thought better of it and taken off.

"Hey, man, I think he's gone. We're not going to catch him."

The unfortunate reality of these situations was that even with the trucks supporting us, in dense, urban terrain, we were usually much slower than the guys running from us. Ideally when clearing an area like this, we would have forces in place to control "squirters" as the enemy forces running from us were called, but the ruins were extensive, and the platoon didn't have the manpower to lock the whole area down without bringing in a second platoon, which would limit our ability to patrol elsewhere.

The rest of the clearing operation took another two hours. By the end of it, we'd found shell casings and food trash that showed us where the shooters had been, but we didn't catch or kill them. When we reached the northern end of the ruins, we linked back up with the trucks, climbed in, and drove to COP Amanche to debrief. The drive was just a few minutes, and after we pulled into the COP, Rob, SFC Murray, and I walked into the TOC while the rest of the platoon reset the vehicles. We grabbed bottles of water and flopped onto the couches to cool off.

Since this was the first time we'd spoken in person since the firefight, we shared our respective sides of the story. Even in a small gunfight like this one, it's always amazing how much everyone's experiences differ. Because SFC Murray and I had employed no method to mark our location in the ruins (after the initial smoke grenades), Rob thought we were clearing much faster than we were, so he pushed his vehicles further north than was ideal because he was rightfully concerned about potential fratricide. This left gaps in our coverage and was probably how the insurgents squirted. We discussed adjustments the platoon could make to tighten this up in the future, and Rob and SFC Murray went back outside to finish up with their platoon.

I sat on the couch, finishing my water. While it was true that with tighter synchronization, we could have reduced the chances the bad guys would get away, that wasn't the whole problem. The real problem was that Mechanics was just too big an area for our troop to secure. We couldn't keep enough guys on the ground to maintain persistent coverage, so once we left an area, the bad guys could just move back in behind us. Similarly, even with living on COP Amanche, we still weren't close enough to the population. The huge walls surrounding our compound made us unapproachable, so even the people who were inclined to trust us were unlikely to come to the COP to provide information, and when we left sector, anyone who *had* given us information would be at the mercy of the enemy forces living in the neighborhood with them.

The only way to bridge the gap would be to move directly into the neighborhood and cover it 24/7, but we'd need a lot more people to achieve this, and where were those people going to come from?

We weren't the only unit to identify this problem, and as it turned out, with the help of one of the units working down the street, the squadron found a solution.

II

FOB Falcon, Iraq, June, 2007

Lieutenant Danly's office was a strange one, but then again, Lieutenant James Danly was a strange lieutenant. A Yale graduate who was serving as the company artillery officer for the 2-12 Infantry, Danly had converted his lackluster Iraqi barracks room into an office of sorts, complete with a large wooden desk that faced the door, a computer, and an Ancient Greek painting of a ship. He was sitting behind that desk when I knocked on his door and walked into his room, and as I entered, he stood to shake my hand. My immediate impression of Danly was that he was both smart and energetic, and he confirmed both over the course of our conversation.[1]

"Hi, are you Lieutenant Danly?"

"Yes, I am. Call me James."

"Hi James, I'm Dan. It's good to meet you."

"You too, Dan, what can I do for you?"

"Our Squadron is taking over 838, 840 and 842 from your unit, and I heard you guys cracked the nut on counterinsurgency."

Over the next half hour, Danly outlined the operation he had developed to combat the insurgency in his unit's sector of Baghdad. It was a simple but effective plan, and when combined with the adjustments our squadron made to how we were conducting our operations, it changed everything about our effort in Iraq.

Our difficulties in Mechanics hadn't been unique. Across Baghdad, units recognized they didn't have the density of forces necessary to conduct counterinsurgency effectively. To address this problem, as the Surge battalions flowed into theater, the theater

---

1. At the time of writing, James Danly is currently serving as the Commissioner at the Federal Energy Regulatory Commission, so my assessment was apparently correct. It's funny where people end up.

commander prioritized critical areas and realigned forces to increase the troop density in these areas. Doura, the neighborhood just north of Mechanics, was a hotbed of the Sunni insurgency in the heart of Baghdad, and it was consequently identified as one of those critical areas.

The 2-12 Infantry had been working in Doura for months and experienced many of the same problems we had been having in Mechanics. Now the neighborhood was due to be split in half, with 2-12 continuing operations in the western half, and the 1-4 CAV taking over the three eastern neighborhoods, 838, 840 and 842. Our squadron was also responsible for securing the Doura refinery and a large palm grove south of it, but because of its economic value, the former was heavily secured by the Iraqi government, and the latter contained few people and was easy to isolate, so we focused our efforts almost exclusively on the three neighborhoods.[2]

To improve our effectiveness in our newly assigned area, the squadron made changes to how the troops conducted their daily operations, first establishing a new COP in 838, then mandating 24/7 presence in all three of our new neighborhoods. With the reduced size of each sector—838 was only 1200 meters long and 800 meters wide—we'd be able to maintain soldiers on the ground continuously, reducing the enemy's ability to coerce locals, emplace IEDs, or move freely. One of the tenets of counterinsurgency is to catalogue and control the population, so the only thing we were still missing was a plan to focus our efforts and the time to implement them. Danly's operation provided us with just that.

Under the plan, each troop was responsible for conducting and maintaining a full census of its assigned neighborhood. By itself, taking a census wasn't revolutionary; units had been cordoning neighborhoods, meeting the people who lived there, and searching their homes for years. What was different about this effort was the administrative rigor with which the headquarters maintained that census data, and the level of focus the executing units had to apply to update it. One of the main reasons so many cordon-and-search operations failed was poor administrative work. The houses in Baghdad are built on top of one another, and it was often unclear which house was attached to which address, and who exactly lived where. Additionally, intelligence databases and enemy target lists were not always shared or disseminated, so even when

---

2. Our old neighborhood, Mechanics, was assigned to a newly arrived unit in theater, the 2nd Striker Cavalry Regiment (SCR). A much larger unit than the 1-4 CAV, the 2nd SCR was able to put significantly more boots on the ground in that area than we ever could.

a search operation found a bad guy, the searchers often didn't recognize his significance, and let him go. Danly's plan focused on addressing these problems.

Headquarters elements used up-to-date map imagery to build detailed maps of each neighborhood, then units on the ground cross-referenced those maps with photographs of the houses, front doors, and entry gates of each dwelling. During an engagement, every person living in each house was photographed, biometrically enrolled in a theater-wide database, and questioned in a non-threatening manner to determine who he was, where he worked, and what problems he wanted addressed by the government. The process was painstaking, but when complete, it provided an accurate catalogue of who exactly lived in the neighborhood. Armed with this information, units could easily determine who in the neighborhood was wanted for crimes elsewhere, and who were new arrivals. Moreover, *the locals then knew that we knew*, and that knowledge changed their behavior substantially, particularly once we had lived in and patrolled the neighborhoods long enough to recognize everyone by sight.

After that initial meeting, Danly invited me to attend one of his company's daily meetings so I could see how his commander and staff coordinated the census. I was so impressed that I recommended he brief Cook and Crider on the program, and not long after their conversation, census work became the squadron's full-time mission. I didn't see Danly again for the rest of the deployment–he left the 2-12 a few months later to support General Petraeus's staff in the Green Zone–but his contribution to the war stands out in my memory as an example of how outsized one person's impact could be.

While this modified census was easy enough to explain, it turned out to be incredibly difficult to execute. The turnover of Amanche to the striker unit went smoothly, and the establishment of a new COP–Banshee–in 838 wasn't difficult, but the new neighborhoods were every bit as tough as Mechanics had been, and the insurgents were fighting to prevent us from making any progress. As with the rest of Doura, our new neighborhoods were almost entirely Sunni. The residents there received little support from the Iraqi government, and they hated the Iraqi police units assigned to patrol the area. Because the police had so little penetrated the neighborhood, Doura served as a haven for Sunni terrorists operating across Baghdad, and they weren't inclined to give up the area without a fight. Defeating them would take a concerted effort by the squadron to secure the neighborhood, kill or arrest any irreconcilable insurgents, convince the rest of the population we could be trusted, and integrate the entire area into the Iraqi government.

To accomplish this, the squadron permanently assigned one troop to each neighborhood. This ensured that through regular engagement, the soldiers would become intimately familiar with the people in their area and thus increase the effectiveness of their efforts there. Initially, Apache Troop was assigned to 840, Bandit took 838, and Comanche took 842, although after the new Bandit Troop commander experienced some issues in sector, the squadron eventually swapped Bandit's and Comanche's sectors, an arrangement that lasted through the remainder of the deployment.

In Apache Troop, we managed our sector by assigning the platoons daily eight-hour shifts. Before each daytime patrol, the platoon leader came to the TOC, and I handed him a thick stack of paperwork with the day's census assignment. During the patrol, his platoon went house to house, sitting down with the family who lived there, enrolling every member of that family in the military's biometric database, and discussing with them their needs and concerns. At the end of every patrol, the platoon leader brought the stack of paperwork back to the TOC, and Bonilla, King, Maine, and I entered the data into our repository and prepared the next day's assignments. The platoons didn't conduct any census operations at night, for obvious reasons, instead patrolling the streets to prevent anyone from emplacing IEDs or terrorizing the locals.

The effort was a grind, and a dangerous one. In June alone, the squadron was attacked fifty-two times, suffering multiple casualties and one Bandit Troop KIA, Michael Pittman. The attacks didn't start to drop off until August.[3] Every engagement had to be secured with roof-top security positions, and, as often as not, while a platoon was conducting engagements on one street, enemy forces would be emplacing IEDs two streets over that the platoon would hit a few hours later. Throughout all of this, the platoons had to maintain absolute restraint in their engagement with the population. A single accident or overreaction by one of our soldiers could reverberate across the area and alienate the entire neighborhood.

Despite the challenges, the squadron continued to press hard on the census. Nobody was sure it would work, but it represented a structured approach that allowed the platoons to focus their efforts toward a defined objective; it was something we could

3. Sills, Tom. n.d. "The Actions of 1-4 Cavalry in 2007/2008 East Rashid Security District."

sink our teeth into. And in fact, by the time we finished our initial census a few months later, the situation in Doura was substantially improved, and we were able to transition our focus from clearing enemies out of the neighborhood to rebuilding it.

But in June, that transition was a long way off, and nobody knew how things would play out.

Part III

Combat Platoon Leader

12

FOB Falcon, Iraq, June, 2007

A few weeks after taking over the new sector, Lieutenant Colonel Crider determined he needed to make a few personnel changes. The months of combat had taken their toll. Some guys had made serious mistakes, others had suffered injuries and been evacuated, and others had been rattled by events that had happened in sector or casualties their units had taken and needed to take a knee. A few, like Major Manville, were even singled out for promotion and pulled up a rung.[1] The unit wasn't receiving any new lieutenants, though, so the decision was made to rotate folks from the staff and support positions into leadership billets and vice versa. My old friends Mike Castillo and Matt Babiarz finally escaped from the headquarters to take platoons in Apache and Bandit Troops, and I was rotated to Comanche to take over an infantry platoon. I was sad to leave behind the guys I'd trained and fought with since 2006, but I was also very ready to get out of the XO's office and into the fight.

I had about three days to hand over my XO duties to Koky, say my goodbyes to the headquarters crew, and drag my duffle bags two buildings over to my new barracks. I was now the platoon leader for 1st Platoon, Comanche Troop, and I'd stay with the boys of 1st Platoon for the rest of the deployment. 1st Platoon had been run by an eccentric lieutenant since it was stood up in 2006, and the platoon sergeant–SFC "Chaz" Hanzich–was very candid during our first conversation about what he thought about both my predecessor's performance and a cavalry officer (which I was) being put in charge of his platoon. Between that, and some of the friction I'd had with Comanche when we were working together in Mechanics, it seemed that getting the guys to trust me might be an uphill battle.

---

1. He was promoted to the Brigade operations officer and replaced at squadron by Major Paul Callahan.

My first interaction with the platoon was a bit strange. They were on a three-day down cycle after completing a security rotation at a nearby Special Forces compound, and when I showed up at the barracks, the guys were blowing off steam with movies and video games.[2] Chaz walked me through the barracks introducing me to people and showing me where my room was; then we talked for about ten minutes about how the platoon and company were running. At the end of our conversation, he said: "You play HALO, Sir?"

I had never played HALO in my life. Being a child of the '80s though, video games were not totally unfamiliar, and once upon a time I'd been a pretty good Nintendo 64 *Goldeneye* player, so I said, "Yeah, for sure. Let's do it." Chaz and I walked to the barracks common room, where three of the guys from the platoon were crowded around a PX television and an Xbox. With typical GWOT improvisation, the platoon had rigged the building up with a locally procured network that connected Xboxes from across the company together to allow twelve-player HALO games. Chaz handed me a controller, gave me a chair, and announced, "Alright, boys, our new cavalry spy of an LT is here. HALO time, bitches!" We sat down. The first thing I had to do was pick a name. On a whim, I typed in "CAVSPY," and we kicked off the first of many games we'd play over the next twelve months. I was never the platoon champion–that title usually went to UPSETMAGUET–but I held my own well enough (in both the game and its accompanying banter) that I never looked too lame. By the end of the game, everyone was laughing and shouting, and after the last power sword killed the last master chief, we all headed to chow to eat and recount the highs and lows of the competition.

As strange as it sounds, in large part, I attribute our regular HALO games to creating the positive relationship I enjoyed with the team and to the close-knit nature of the platoon in general. Camaraderie was hard to build on a deployment when every guy had his own computer, cellphone, TV, and music, but HALO brought the team together. After that first game, I never felt like an outsider, and by the time we were back on patrol a few days later, running the platoon felt comfortable.

---

2. Assigning conventional infantrymen to guard Special Forces compounds was common during the GWOT. As a Special Forces detachment only has twelve personnel, it is not organically capable of securing itself while also conducting regular operations.

It was a particularly hot and steamy day when we found our first IED. Comanche hadn't taken over 838 from Bandit yet, so today we were conducting a dismounted presence patrol in *Mahala* 842. Our mission was to engage with the population and demonstrate to the people of Baghdad that American forces–and the Iraqi government we represented–were looking out for their well-being. The drive to 842 took about fifteen minutes, and on the way, the intelligence section reported to us that they had heard an IED had been emplaced along the route we took into sector. Preferring to find the IED the good way rather than the bad one, I decided to stop short, dismount, and conduct a foot patrol through the high grass south of Route Senators to see if we could locate the device. We took a detour and found a back road that paralleled Senators, driving down the road until we were near the suspected IED location. We stopped the Humvees, and third squad and I got out to sweep the area while the rest of the platoon overwatched our movement from the vehicles.

The area we were searching through was adjacent to the palm grove in the eastern portion of the squadron's sector. The ground was muddy, and entirely covered in six-foot tall grass. The air was thick with humidity and flies, and we couldn't see more than a few feet. Sergeant McDowell and I took point, the squad fanned out in a V-shaped formation behind us, and we began pushing our way through the tall grass looking for the bomb. My plan was to sweep the area between the back road and Route Senators to determine whether our route into sector was clear.

One of the challenges of being a platoon leader is trying to maintain the image of confidence you know you're supposed to project to your guys when you have no idea what to do with the situation at hand. To this point in my career, I had neither patrolled in tall grass, nor had much experience looking for IEDs. We'd found a few in Afghanistan back in 2003, but they had been anti-tank mines buried in a Tatooine-esque wasteland, which was a good bit different than this situation. As I pushed through the grass, I wondered what the bomb we were looking for might look like–maybe a big, black orb with a fuse sticking out of the top? I knew at least what to do when we found it–call the explosives disposal guys–but the rest of the approach to dealing with IEDs was at this point beyond me.

Progress was slow as we forced our way through the grass. An hour into the movement, our uniforms were soaked with sweat, and as we neared the steep embankment that led up to Senators, I was just thinking about calling a halt so the trucks could pick us up when I saw a piece of wire hung in the grass ahead of me.

"Hold up, I found something."

McDowell chimed in, "Yeah, me too–there's a piece of wire here."

A couple of the other guys added that they'd found it too; there was a hair-thin piece of stereo wire strung out across the direction we were walking, so the entire patrol had found it more or less simultaneously. The wire led from the general direction of the palm groves to the embankment that paralleled Senators. Even to a guy as fresh on the ground as me, it was about as plain an indicator as there could be that there was probably an IED in the area.

McDowell said, "What now, sir?"

"What now, sir?" The timeless question posed to people in charge. My mind was blank for a moment. I remembered the various slide shows I had seen on IEDs, and I remembered talking with the Apache guys who'd found plenty of them over the last few months, but none of that captured the actuality of this moment. I knew we needed to do something with the bomb, but we hadn't found a bomb. We had found a wire, and I didn't know what to do. I certainly didn't want to look like an idiot calling EOD if this turned out to be nothing.

"Let's see where it goes."

I picked up the piece of wire, reoriented the patrol along it, and we walked along the wire to find the bomb. In hindsight, this was an idiotic decision. Why would I pick up the wire and push blindly through the grass *toward an object I expected to explode?* Clearly, the wire indicated a bomb was in the area, and if I had been thinking a bit more intelligently, I would have realized it also indicated there was probably someone at the other end of the wire waiting for the right time to detonate it.[3] As we traced the wire through the grass, some of these thoughts crossed my mind, but as was so often the case, I didn't realize how much better I could have handled the situation until it was too late to matter.

About a second later, I heard, "Hey sir, it looks like the wire is stuck on something."

McDowell and I shook the wire a bit to see if it was caught in the grass. The wire didn't come free, so he reached ahead to push aside an armful of grass and see what the issue was. Of course, as anyone with any sense already knows, there was an explosive sitting two feet in front of our faces.

---

3. This type of IED is called a command detonated IED. It is comprised of an explosive (generally hidden) with a wire leading from the explosive to a spot where the operator can watch the site with a detonator in hand, waiting for the right moment to set the device off and blow his enemy to bits.

"Shit," I remarked. "That's an IED. Back up."

The IED looked as cliché as can be imagined. It was about the size of a small Federal Express flat rate box, and it was wrapped in black plastic and electrical tape. An electronic doodad was perched on top like a bow on a Christmas gift. The whole thing was lying in the grass about 15 feet from the road, and if it blew up, it would probably kill us all. We all pushed back a bit and gathered in a circle about twenty feet from the IED.

We stood there for a moment. Then, after a few seconds of looking at each other, I don't remember who it was, but somebody started laughing. Then another guy started, and then we were all laughing at the ridiculousness of the situation. Here we were, a group of eighteen-to-twenty-seven-year-olds, walking through a farmer's overgrown field looking for a bomb we heard about from some Iraqi on the tip-line–*maybe even the same Iraqi who had emplaced the bomb and was looking to set a trap for us*–and we were doing it the dumbest way imaginable. It called to mind childhood images of Wile E. Coyote myopically following the string of one of his own crazy contraptions until he fell off a cliff.

The moment passed, we collected ourselves, and I called up Chaz on the radio.

"Comanche seven, this is Comanche six. We have PID of an IED at my location. We'll establish security here. Mind requesting EOD?"

"Roger, working it," he replied.

I couldn't hear his conversation with the company headquarters, but I knew Chaz would work out the details. Our trucks had two radios. As the platoon sergeant, Chaz always kept one on our internal frequency, which the dismounted element shared, and one on the company frequency so he could talk to our headquarters back at Falcon. Everyone else in the platoon kept one radio on our platoon frequency and the other on a squad internal network so they could coordinate internally, bullshit, and talk trash to each other without anyone else being able to hear it.[4] While he called up the request for EOD support, the rest of us fanned out in a large security perimeter around the IED and waited.

EOD took about an hour to arrive. Waiting for EOD to arrive was a hallmark of my time in Iraq. Because the number of IEDs found in any given day across the theater far outweighed the number of EOD teams available, you could easily wait one, two or even three hours for a team to show up and take care of your IED. It was like being on hold with your insurance company after you got into a fender bender. You needed to talk to the insurance company *now*, so you couldn't really call back later, but because you were on the phone, you really couldn't do anything else until the agent answered. Waiting for EOD was like that, except that we didn't have air conditioning, everyone involved had guns, and the other driver's friends might fire mortar rounds at us or make his car explode. Eventually, we got pretty good at occupying ourselves during the wait, but early on, when we still had the impression that every minute we spent waiting was one the enemy was using to fortify his position or bury IEDs or something, waiting on EOD could be quite frustrating.

After the longest hour in the history of recorded time, we finally heard the EOD team's traffic on our platoon frequency that heralded its arrival.[5] The EOD trucks were shockingly large. Called buffalos, they look like enormous, armored Suburbans that had been freshly driven off the set of a post-apocalyptic movie and equipped with armored firing portals, crane arms, and heavy hydraulic doors and hatches. Their convoy pulled up next to our trucks, the imperial shuttle-like rear staircase of the buffalo hissed open, and the team leader climbed out. Chaz and I walked over to talk to him.

"Hey man, how are you?"

"Good. We're here to take care of the IED."

"Alright, c'mon, I'll show you where it is."

We walked the hundred yards or so toward the IED in silence. When we got to the spot where McDowell and I had stopped, I pushed the grass aside and pointed at the IED, which was once again about two feet away from us.

"There's the bomb. That copper wire leads off toward the palm grove."

The EOD guy's reaction was immediate. He backed up about ten feet, turned around, and started walking quickly back to the trucks. A bit confused, I caught up with him and asked him what was up.

---

5. A standard technique in theater was for any support element arriving to assist a ground force to switch a radio to the ground force's working frequency to coordinate the details of the required support. This was necessary to ensure the arriving support didn't do something undesirable, like accidentally drive across the IED themselves or shoot at dismounted, friendly security forces.

"Aren't you going to look at the IED?"

"No, that's not how we do this. I thought you were walking me over to the general area of the IED, not leading me directly up to the damn thing! It's still armed! What if it had blown up?"

"Huh? It didn't blow up earlier. Why would it blow up now?"

The EOD team leader was clearly irritated with my question. The contrast between my platoon's attitude and the EOD teams' attitudes toward IEDs remained stark throughout that deployment. I won't generalize and say it's a community-wide cultural difference, because there are infantry guys who take IEDs very seriously, and there are EOD guys who don't, but I can confidently say that for better or for worse, we were always much more casual in our dealings with the IEDs we found than the EOD guys were. Even by the end of the deployment, when we were experienced with dealing with IEDs and had certainly had our share of them blow up on us, for some reason, when one was sitting in front of me, a sort of fatalism took over, and I never had the sense my life was on the line.

This disparity–and our resultant feelings regarding the time we felt appropriate to spend on reducing a given IED–regularly caused friction between our organizations. EOD guys always seemed to feel that caution was the best course of action, while our priority was generally to take care of the problem quickly enough that the chow hall would still be open by the time we got back to base.

"We do this for a living. Secure the area, and I'll send over the robot."

"Umm. Okay. A robot. Sure. Yeah, okay, let me know what you need from us."

The team leader went back to his truck and climbed in. I walked back to Chaz's truck to wait. Nothing happened for a long time, so naturally, we started speculating about what was going on in the truck.

"Do you think they have to brief the robot on what it's about to do?"

"No idea. Maybe they give it a motivational speech or a pre-game massage or something."

Minutes passed.

"How do you think the robot feels about its lot in life?"

"Dunno. He's got a pretty shitty gig. Nobody else is willing to take care of the bomb, so he's got to do it himself."

"Do you think the EOD guys get the valor awards, or do the robots get them instead?"

"Do they send a letter to the robot's family when he gets blown up?"

"Is there a home for retired robot veterans where they can live out their robot lives watching robot *Wheel of Fortune* and sexually harassing robot nurses?"

More minutes passed.

"What do you think is for chow tonight?"

"I'm pretty sure it's stir-fry day."

"Awesome. I hope we make it back in time."

"What the fuck *are* they doing in that truck?"

"I don't know man, but they need to hurry the fuck up."

EOD did not hurry the fuck up, and many more minutes passed before the rear of the buffalo hissed open again, and the robot rolled out and parked near where we were standing. This was my first experience with an EOD robot. It looked something like Johnny-Five (the robot from the '80s movie *Short Circuit)* except that it had only one arm and was tethered to the EOD truck by a long cable trailing out of its rear.[6] The robot drove off the ramp and stopped. It was holding a block of plastic explosives in its hand. The EOD team leader was nowhere to be seen.

Chaz and I stood there looking blankly at the robot for about thirty seconds. I'm not sure what I expected to happen. I think I was waiting for a cool, digitized version of the EOD team leader's voice to emerge from the robot somewhere, or at least for him to get out of the truck and come talk to us again, but neither happened. Instead, without warning, the robot suddenly lurched off toward the IED, entered the tall grass, hit an uneven patch of ground, and promptly fell over. As its treads spun helplessly backward and forward attempting to right itself, Chaz pointed at it and laughed derisively.

"What a piece of shit!"

We looked from the wallowing robot to the EOD truck and back to the robot again. Minutes passed.

"Do we flip it back over or something?"

"I don't know, man. I don't know."

"At this rate, we're definitely missing chow. Is X-ray going to get us to-go plates?"

"Let me go check."

Chaz walked off to call headquarters, and I stood there watching the robot. While its humming treads spun helplessly backward and forward, I picked up my radio and

---

6. Incidentally, the robot was not (to my knowledge) alive; it did not have a charming, '80s synthesizer voice; and the operator didn't look anything like either Steve Gutenberg or Fisher Stevens.

described the goings on to the platoon on our internal frequency.

"All Comanche-one elements, this is Comanche one-six. Update follows. EOD is on site and working on the IED. They have deployed a robot with C4 to BIP [blow in place] the IED, but that robot has currently fallen over. Maintain the perimeter."

"One-one, roger."

"One-two, roger."

"This is one-three, roger. Is X-ray going to get us to-go plates?"

Chaz chimed in. "This is Comanche seven. I'm working it now."

After some time, the buffalo's ramp hissed open again, and an EOD guy walked hurriedly over to the robot, righted it, and walked back to the truck. The robot hissed to life again and pushed its way through the grass. It disappeared from my view in a minute or so, although between the cable that trailed after it and the wobble of the tops of the grass stalks in its path, I could still detect its whereabouts. The robot must have eventually found the IED because the grass and the cable stopped moving. About thirty more minutes passed, and nothing happened. The brief period of excitement surrounding the deployment of the robot passed, and once again, the area lapsed into a sweltering silence. Chaz walked back up.

"They tell you what's up?"

"Nope. You hear anything on the radio?"

"Nope."

"Let's walk over there and ask them what's going on."

We set off toward the buffalo to talk to the EOD team. About halfway there, a thunderous explosion rocked the area, and Chaz and I took a quick knee to figure out what had happened.

"What the fuck was that?"

"Was it them blowing the IED or indirect?"

"It wasn't indirect. There was no whistle. Couldn't have been an RPG either."

"I guess it was the IED then. Nice of them to tell us before they BIP'd it."

"All Comanche elements, this is one-six. Everyone okay?"

"Roger."

"Roger."

"Roger."

"Alright. seven and I will figure out what happened."

We finished the walk over to the buffalo and waited behind the vehicle. Buffalos are designed to prevent exterior access by enemy forces in urban areas. The bottom edge of the doors is around four feet off the ground, the windows are tiny and armored, and the rear ramp isn't accessible from the outside when closed. These features ensure terrorists (or kids looking for candy) can't easily get at the crew of the truck. They also ensure it's difficult to see who is in the truck or what they are doing. As Chaz and I stood outside the truck waiting to be acknowledged, we again speculated about what the EOD guys might be up to.

"Do you think they wear pants in there?"

"Probably not. I wouldn't. It's hot as hell, and nobody can see what you're doing in there, so why bother? They could be having a gang bang in the back of that thing, and even if Sergeant Major Champagne himself were standing six feet from the truck, he'd have no more idea about what's going on in there than we do now."[7]

"That's true."

Personally, I figured they were probably just playing Game Boy in there or something, but maybe Chaz was right. I'll never know. Eventually, one of the rear portals in the buffalo opened a crack, and the team leader's voice emerged.

"What?"

Chaz and I looked at each other.

"What do you mean, 'what?' Was that explosion you guys? Is the IED gone?"

"Oh. Yeah, we're recovering the robot now. We'll be off target in fifteen mikes."

Chaz laughed and shook his head. My blood pressure increased by a few points, and I lashed out.

"Hey, shithead, did it occur to you that we might be interested in knowing about the explosion before it happened? I've got guys on the ground in a perimeter around the site."

There was no response from the window, and a second later, it shut. Chaz and I stood there for a few more minutes, then gave up and walked back to our trucks. The buffalo not only made its occupants virtually immune to small arms fire, it also made them unassailable by irritated infantrymen. Moreover, because EOD teams were assigned rotationally, the odds of us ever seeing this team again were negligible. Even if we'd tried to find them or their chain of command on FOB Falcon, it would have

---

7. Command Sergeant Major Champagne was the Brigade Command Sergeant Major of the 4th Brigade. He was a hard-nosed infantry guy and the resident enforcer of standards and discipline. Woe to the platoon sergeant whose guys didn't have their gloves or eye protection on when the CSM showed up for a surprise visit.

cost Chaz and me both of the precious hours of time we had between returning to base and bed, and the EOD leadership probably wouldn't have cared that much, anyway. Nobody was hurt, so it wasn't worth the effort–especially on stir-fry night.

"What a bunch of douchebags."

"Yeah, turds for sure."

I got back on the radio. "All Comanche one elements, this is Comanche one-six. BIP complete. Collapse the perimeter, mount up, and prepare to continue mission."

The day wasn't getting any younger, the temperature outside was already over a hundred, and we still needed to get to 842 for our dismounted patrol. As I wrenched open the ever-stuck Humvee door and sat down, I savored the rush of cool air. The feeling of getting back in the truck after a dismounted action of any sort was always amazing, at least when the truck's air conditioning worked, and this truck's A/C was one of the best in the platoon.[8] The driver of McDowell's truck, Eubanks, had a heat injury once and was sensitive to getting too warm, so had rigged up the truck's A/C to pump through a set of long hoses that ran into each of the seats. When nobody was in the back seats, one of these hoses was angled to blow directly on a box of bottled water to keep it cool. When someone was in the seat, the hose could instead be crammed into the front of his armor, up a sleeve, or into the crotch-hole of his pants to cool him down.[9] I grabbed a bottle of water, crammed the A/C hose into my pants, put on a vehicle headset, and listened as Chaz reported the completion of EOD's mission to our troop headquarters. When Chaz was done, I keyed the mike and directed the platoon to drive to 842.

The movement was short–not even long enough for the sweat that had soaked

8. The maintenance of a truck's air conditioning system was always a primary concern of my guys. Each team of four had a HMMWV. The maintenance of the team's vehicle fell on the driver's shoulders, and as he knew he would spend most of his waking hours in the vehicle, the driver was highly motivated to keep the air conditioner in good shape. A driver's reputation in our platoon was in part built on his ability to scrounge parts, sweet-talk mechanics, or otherwise figure out how to keep the interior temperature of his HMMWV at a tolerable level.

9. Through a bug in either their design or production, the Army's issued uniform pants were weak in the crotch. The pants were notorious for tearing across the crotch seam when guys kneeled, high-stepped over something, or got out of a vehicle, so no pair that was worn for more than a month or so had an intact crotch. This occasionally led to an inadvertently racy photo.

through my clothes to dry–and while we drove, I issued guidance to the platoon on the upcoming patrol.

"All elements, this is Comanche one-six. Once we get to sector, one-one and one-three will dismount with me at the south end of town and hit Main Street. One-two and seven will rove around sector to maintain presence everywhere else."

"one-one, roger"

"one-two, roger"

"one-three, roger"

We pulled into *Mahala* 842, stopped our trucks on the main street in town, and dismounted. 842 was a bustling, commercial section of Doura with few homes but many businesses, and the street we were standing on was packed with Iraq people going about their daily life. On the corner next to our lead truck, two men in *dishdashas* haggled over the price of a skinned, fly-covered goat carcass hanging outside a butcher's shop. Across the street, a cluster of chatting women in *hijabs* led their small children toward the *Al Nabaha* school. A stooped, white-haired man walked past us, winding his way through the crowd with a wheelbarrow full of live fish from the Tigris and clanging a bell to advertise his wares. 842 was located just a few hundred yards from the massive Doura refinery, so the burning-tire smell it produced mingled with the usual Baghdad smells of food, exhaust, sewage, and garbage to yield an almost tangible, aromatic haze.

One-one's squad leader, Tucker, and team leader, Blanco, met me near the lead truck with one-three's leadership, Van Awesome and McDowell. The platoon's interpreter, Ozzie, joined us as well. Tuck's, Blanco's, and Ozzie's uniforms were still the light-gray color of sweat-free uniforms that hadn't been out of their trucks yet today. The rest of us were dark-gray and dripping. The crowd of Iraqis walked around us without paying us much mind, except for a couple of children, who approached us to beg for candy with cries of "Mistah! Mistah!" This time, the kids were out of luck, and Blanco sent them away with a vigorous hand gesture and a firm, "*za!*"[10]

"Alright, let's push north and make friends. No specific agenda or meetings today, we're just here to show people we're here. Tuck, you guys take the far side of the road, Van, you take this one. I'll bounce between you with Ozzie. Shoot me a call if you need him to talk to someone."

The guys nodded in assent and walked back to their trucks to corral their guys.

---

10. Arabic slang for "Go!"

Because a Humvee took at least two people to crew–a driver and a gunner–unless we left a truck at the FOB, the squads usually had only one or two bodies for dismounted patrolling. Today, 1st Squad had only Tuck, Blanco, and our communications guy, Smith, while 3rd Squad had Van Awesome, McDowell, and Sproul on the ground. We fanned out and started walking slowly north, nodding to the locals who made eye contact or greeting them if they seemed friendly. A few months later, when we knew everyone in town, we'd be inundated with requests or complaints before we'd walked more than a block. Today, though, none of the locals had much to say to us, so we made our way along the street with the trucks idling along between our teams, scanning for terrorists and taking in the local sights, sounds, and smells.

We walked past a mechanic shop, where a Daewoo Prince was pulled onto a set of concrete ramps next to a stack of old tires in front of the shop's sliding, orange, aluminum door. The mechanic had an oil filter and a rag in his hands, and he was draining the oil out of the bottom of the car and into the gutter. I met his neutral gaze for a few seconds and continued walking, thinking to myself that he was dressed exactly like Billy Joel in the video for *Uptown Girl.*

"He's a got a hell of a better tan though, for sure."

"What's up, LT?" asked Ozzie, looking over at me.

"Nothing, man, just talking to myself. Want to talk to anyone?"

"Let me buy a soda real quick."

"Sure, let's do it."

Ozzie and I walked across the street to a convenience store. The store was a shallow inlet in the wall of one-story buildings that lined the street. It had an open storefront that consisted of a cooler rack, behind which sat a guy with short black hair, a neatly trimmed mustache, and a ready, white smile. Ozzie negotiated with him for a cold Fanta, and he and the man exchanged pleasantries in Arabic for a minute or so.

"Hey Ozzie, ask him if he's had any problem with terrorists lately."

Ozzie nodded and rattled off a string of Arabic to the shop owner. The owner's brows knitted briefly, then he began talking to Ozzie animatedly, emphasizing his words with hand gestures. He talked for more than a minute, then stopped. Ozzie nodded, turned to me, and said, "Nah, he says everything's fine."

A minute-plus soliloquy by an Iraqi in Arabic had become a five-word sentence in English. "There's no way," I thought, "that all he said was 'everything's fine.' What were the other ninety seconds about?" This was very common when working with Ozzie. At times, I was bothered enough to press for more information, but since this was rarely productive, I didn't do it often. Ozzie was a good interpreter, but he was still a human, and he got hot and irritated like everyone else. Also, his daily job consisted of mostly asking the same two-dozen questions to people and trying to look interested in their responses, complaints, and criticisms, so he got as bored with the process as the rest of us did. His situation was further complicated by the fact that he was an Iraqi, which meant he caught an extra ration of shit from some locals who perceived his employment with us as disloyalty to his people. This never seemed to bother Ozzie much, but it must have rankled sometimes.

"Okay, Ozzie, thanks." I turned to the shopkeeper, said "*Shukran,*" and headed back toward the main street to continue the patrol.[11] While Ozzie and I had been discussing terrorists and buying Fanta, Blanco had struck up a conversation with an old man, and Blanco gestured to me to request Ozzie's assistance. Some of my fellow officers liked to be involved in every conversation their guys had with Iraqis, but I was happy to let my guys take the lead sometimes. I sent Ozzie to Blanco and walked over to talk to Tuck instead.

"How're things, Tuck?"

"Good, Sir. Just keeping an eye out for CHUDs."

CHUDs, or, Cannibalistic Humanoid Underground Dwellers, were always a topic of conversation in first squad. Guys found a wide variety of ways to occupy their brains in Iraq, and Tuck's was CHUD hunting.[12]

"You see that manhole, sir? The sewer under there is probably full of CHUDs. Mostly, they come out at night. Mostly."

Tuck had an encyclopedic knowledge of bad '80s movies. Being a fan myself, we often discussed them on patrols. Who was I to say there weren't CHUDs in the Baghdad sewers? All the oil and chemical runoff might have made any number of crazy things grow down there. Chaz had a wild idea that we should all load up with vaccines and other prophylactics and lead a patrol through the sewers to rule out the possibility once and for all, but the only time we ever actually saw anyone enter one was when Blanco accidentally dropped a full magazine down an open manhole and paid a kid twenty bucks to recover it. That kid wasn't too pretty to look at when he came out, so the talk

11. *Shukran* is Arabic for "thank you."

12. CHUDs were made famous in the appropriately named 1984 film, C.H.U.D.

of patrolling the sewers ceased.

"You're probably right, Tuck. Maybe the CHUDs put in that IED earlier."

"Nah, that's not their way. Also, that spot was too far from any sewer entrances."

I nodded in agreement. Tuck was probably right. There was no way CHUDs put in that IED. Life continued for the people of 842 as Tuck and I mused about CHUD behavior for a few more minutes while Blanco and Ozzie wrapped up the conversation with the old man. When they were finished talking, Blanco and Ozzie walked over to fill us in on how the conversation went.

"Hey sir, that old guy was pissed off because the Iraqi police at the checkpoint took his ration card, so now he can't get propane anymore."

As I'd found out in Mechanics with Apache, losing your ration card was a big deal, as it meant the old man would not be able to draw any propane until he got a replacement card. Because of their value, ration cards were a hot black-market commodity. Assuming the old man was telling the truth, that's probably why the police officer had taken it in the first place.

"Which police checkpoint took it, and what did you tell him?"

"He said they took it when he was stopped at a checkpoint in Karada. I tried to get him to show me where it was on a map, but you know how Iraqis are with maps. I couldn't get a solid answer, so I told him we'd report it and see what we could do."

Karada was not far north of us, but it was in another brigade's area of operations, and that brigade was based somewhere in the Green Zone. I didn't know it at the time, but the odds of me finding the U.S. unit which owned the area, convincing that unit to put me in contact with its Iraqi Police counterparts, figuring out which checkpoint might have taken the old man's card, and then identifying which police officer at that checkpoint did it, were dreadful. Later in the deployment, I would have just taken the old man to Dr. Manza and gotten him to give the old man some propane on the side, but I didn't have that as an option at this point, and I still believed these sorts of problems were solvable through the proper channels.

"I'll report it up. Goddamn police. The people here hate them, just like they did in Mechanics."

"Can't blame them, really. They're Sunnis. The police are Shi'a. They all hate each other."

"Yeah. Alright, let's keep moving."

We walked another four or five blocks to the north end of the *mahala*. From here, we could clearly see (and smell) the colossal smokestacks of the refinery north of us. The dismounted patrol had taken us about three hours, on top of the four hours or so we spent on the IED. I was hot and hungry, and my neck and back were stiff from the helmet and armor. I waved over Van Awesome and McDowell, and we circled up with Tuck and Blanco to talk.

"Van, did you and McDowell see or hear anything interesting?"

"Nope. Nothing. I thought I saw Osama Bin Laden once, but it turned out to be two dogs fucking."

"Damn. That would have been a good find."

Everyone nodded and fidgeted a bit adjusting their armor.

"All right, let's move back to the trucks and head back to the FOB. I think we met the intent of the patrol."

We loaded up and drove back to FOB Falcon without incident. On the drive, the radio echoed with bullshit and laughter as the guys in the platoon took shots at one another and cracked jokes. The guys were generally in a good mood on the way back to base, and today was no exception. When we got back to Falcon, we stopped at the gate to dry fire our weapons into sandbag-filled clearing barrels as a final safety check before entering the base. The drivers then dropped us off at the barracks and headed to the wash racks to refuel and clean the trucks, while the gunners cleaned all the heavy weapons and secured the ammunition. Chaz and the squad leaders talked through personnel issues and conducted a quick debrief with the First Sergeant, and I walked up to the headquarters to peck out my daily report on Tigernet.[13]

With all my administrative work complete, I found Chaz, and we headed to chow.

"Look at that, we even made it back in time for stir-fry night."

---

13. Tigernet was one of the seemingly infinite number of systems the Department of Defense commissioned during the GWOT in its quest to visually depict and deconflict what the hell was going on during combat. Tigernet was a system that allowed leaders to tag events and photos to locations on a map interface. After completing a mission, every leader across Iraq sat down at his terminal and typed out a summary of what happened on the patrol. As an end user, I never knew if what I typed into the system mattered much. As the months rolled by, and my entries varied from boring, to comical, to seemingly strategically important, I began to suspect nobody read them at all, but I could have been wrong.

13

Mahala 838, Iraq, July, 2007

The first glass of Fanta I drank every morning was cool and refreshing. The second and third were fine, and the fourth was manageable, but by the seventh or eighth, it became difficult to accept the offered glass with a genuine smile and an enthusiastic sip. Fanta was a critical part of getting people to trust us though, and by consuming it cheerfully, I hoped to eventually reduce the number of IEDs our platoon hit.

Comanche Troop had been running *mahala* 838, known locally as *Abu Tayara* (Airplane) because of the large airplane statue at the north end of the neighborhood, for several weeks under our new commander, Captain Bret Hamilton. The Troop's three platoons, 1st Platoon (mine and Chaz's), 2nd Platoon, led by Lieutenant Rich Smith and Sergeant First Class Jason Lewis, and the mortar platoon led by Lieutenant Jason Fedish and Sergeant First Class Stephen Hendrix, rotated through patrolling the *mahala* in eight-hour shifts. The troop ran a swing shift, each of the platoons doing three days of mornings, three days of afternoons, and three days of security at COP Banshee. Bandit Troop, which had experienced some leadership issues, provided the full-time night shift for both our sector and Apache's.

Unless we were attacked or hit an IED, we generally spent our eight hours in sector conducting the day's census assignment, going door to door along our assigned streets and engaging with the people we met there. We had three squads and seven trucks, so I took one squad and two trucks with me for the census work, while the other five trucks split up to patrol the rest of the neighborhood. The census mission was considered the duller of the two options, since while the squad leader and I were talking to locals with Ozzie, the rest of the section just sat outside the house pulling security, so I rotated the three squads through this duty to keep them from getting too bored. Today I was with 2nd Squad, which was led by Staff Sergeant Fox and Sergeant Gonzales. Fox was an old school infantry NCO. He was hardworking and organized, and he ran a tight squad, but he had no interest in talking to Iraqis about their problems and preferred to manage security around the engagement rather than participate in it. Consequently, Gonzales–or Gonzo as he was called by everyone in the platoon–accompanied me and Ozzie to our meetings. Gonzo was a quiet, friendly guy with a fatherly streak. He ran a good team, but he also liked the kids in the neighborhood and enjoyed engaging with the people. He even liked Fanta, which was a good thing since the Iraqi lady whose gate we were standing at was offering us a tray with three glasses of it.

I smiled and took a glass, saying "*Shukran*" as I drank the orange soda with a smile. Gonzo did the same. The old woman's face broke into a huge grin. The old women in our neighborhood always insisted that we have something to eat or drink during our engagements, and they always seemed delighted when we enjoyed it.

"Ozzie, would you mind giving her our usual introduction?"

"No problem LT." Ozzie spent a few minutes describing to the woman and her equally old husband that we would like to come in for twenty minutes or so to ask them a few questions. Ozzie was an engaging guy, and even though he was Shi'a, he got along well with almost everyone in 838. A platoon's success in this sort of mission was heavily reliant on its interpreter's ability to engage with people, and Ozzie was one of the best in the squadron. Some interpreters wore ski masks to disguise their appearance and avoid the reprisal that being recognized risked, but Ozzie never did, and the people in our neighborhood appreciated his open, friendly demeanor.

When Ozzie finished his introduction, the old man nodded and gestured for Gonzo, Ozzie, and me to follow him into the house. His wife didn't come with us. Women rarely took part in our engagements. We followed the old man into his living room, introduced ourselves with the right hand on-heart gesture commonly used in Iraq, and sat down on a couch that groaned under the weight of our fully-kitted-out 260-to-300-pound frames.[1] Gonzo pulled out his notebook to record the necessary information from the meeting. We usually rotated who led the conversation, with the other guy taking notes.

I began the meeting. "Thank you for welcoming us into your home today. We're going door to door across the entire neighborhood to meet the people of *Abu Tayara* and to get to know you. We want to get rid of the terrorists ruining life for the people

---

1. Even though it was probably off-putting, we always conducted our meetings in armor and with weapons. The gear wouldn't help us if the house exploded, but wearing it ensured that if we had to get out of the house quickly to respond to something in sector, we could do so.

here and help rebuild the neighborhood, but to do this, we need to understand what problems you have." I paused for a minute to let Ozzie translate, then continued. "Would you mind telling me your name and occupation?"

Through Ozzie, the old man replied. "My name is Hamed Mustafah Sayid Al-Dulami, and I was an electrical engineer under the old regime. Now I am retired."

Gonzo wrote this down as I continued. "Wow, an engineer. That's great. We've got a lot of electrical problems in the neighborhood."

We *did* have a lot of electrical problems. The Baghdad electrical grid only pushed a few hours of power to 838 a day, and even that service was frequently interrupted. I suspected this was for sectarian reasons, but whether that was true or not, most of our neighborhood had no electricity most of the time. A few enterprising locals had compensated by installing huge diesel generators at the northern and southern ends of *Abu Tayara,* and then running stereo wire from the generators directly to any house in the neighborhood willing to pay for the service. The result was a rat's nest of electrical wire wrapped around every electrical pole and running across every street in 838. This *ad hoc* electrical network was both dangerous and inefficient. The wiring used was low gauge and poorly insulated, so not only did the residents lose a good percentage of the power they paid for during transmission, but they risked electrocution when a live wire broke and dangled in the street. The generators themselves were poorly maintained, smoke-belching disasters, but they provided the only reliable source of power in 838.[2]

With this problem in mind, I questioned Hamed about his thoughts on the electrical problems in the neighborhood. He gave me a long answer detailing the problems with the local grid. Ozzie, Gonzo, and I had zero technical knowledge about electrical systems, so Hamed's answer was mostly lost on us. Still, it sounded credible, so I asked a follow-up question.

"He really seems to understand the problem. Would he be willing to help us get the power running? We can fund the material costs."

Ozzie translated, then Hamed shook his head.

"He says it's not his job."

"Whose job is it then?"

"He says it's the *Belladia's* [the local municipal building] job to fix it."

"But the *Belladia* hasn't fixed it, and they probably aren't going to fix it. I'll pay him

---

2. That is, until one of them caught fire and burned completely to the ground later in July.

to fix it, and then he'll have electricity himself, and so will the rest of the neighborhood."

Again, Hamed shook his head.

"He says it's not his job, LT."

"Then he won't have electricity. Is he okay with that?"

"Inshallah." I didn't need Ozzie to translate this for me. *Inshallah* was a word I heard all the time. It literally translates as "God willing" or "If God wills it," and it was used in all sorts of situations. Sometimes, it meant the English equivalent of "I hope that's what happens." Other times, it meant, "I guess we'll see." Still other times, it seemed to mean, "I don't give a shit." Over the course of the deployment, I came to associate *Inshallah* with the fatalistic resignation that we commonly encountered in our engagements with the Iraqi people. At the time, it was immensely frustrating. Why did I care more than the people who lived here about their problems? Looking back on the situation, though, my perspective has changed.

The Sunni population of *Abu Tayara* had a much more realistic perspective on the Iraqi government's willingness to help them solve their problems than I did. To me, a government was essentially an objective entity that provided services to the people on a contractual basis, regardless of their beliefs or ethnicity. This was naïve. In Iraq, the denial of service by the Shi'a government to the Sunni population was both retribution for decades of that same treatment in reverse and a lever to ensure the Sunni population remained the underdog in the newly democratic country. The Iraqi people of 838 understood this, but I did not.

"Alright, I guess he'll just sit here and sweat his ass off then," I thought. Turning to Ozzie, I continued the meeting.

"Aside from electricity, what problems does he have that we can help out with?"

Hamed's response to this was long, and it took a few minutes for Ozzie to receive and process it. When Hamed finished, Ozzie turned to me and said, "He says there are bad people living in the neighborhood that we need to arrest. They aren't from here, but they've moved into the abandoned houses and are threatening locals." When Ozzie paused, Hamed pulled out a small envelope and handed it to me. Inside, were two AK-47 shell casings and a printed pamphlet.

"What's this say, Ozzie?"

He read the pamphlet and said, "It's a death threat, LT. It's from some group that says they'll kill this guy if he doesn't leave the neighborhood. The shells are a traditional Iraqi threat."

"Which houses are the bad guys in? We'll get rid of them."

Ozzie translated, and Hamed shook his head, saying, "No, don't do that. They'll know it was me who told you about them, and their relatives will kill me."

I thought about this for a minute and said, "How about you show us which houses they're in, and we'll keep going house to house like we are right now, but when we get to their houses, we'll arrest them."

Hamed nodded to this, and replied, "Yes, that will work. You must be careful, though, or they will kill me."

I agreed to be careful. Hamed told us the street number and address of the terrorists. I never had much luck with getting Iraqis to reference specific houses on the map I carried, but after a bit of back and forth, we got a house number, gate description, and the physical appearance of the two bad guys. The house was one block from Hamed's house, so we wouldn't get to it today. 2nd Platoon would work the population engagements there tomorrow, though, so I made a mental note to talk to Lieutenant Smith about it and finished up with Hamed.

"Thanks, Hamed, we'll get these guys off the streets by tomorrow."

"*Inshallah*"

I smiled, and all three of us stood up, exchanging farewells as Gonzo and I left the house and walked back to the street. On the way, I asked Gonzo what he thought about the whole conversation.

"It was good, sir. I hadn't thought about the census as an opportunity for intelligence collection, but it seems like it's got a lot of potential. When we're sitting in their living rooms, the people feel safe, and when they feel safe, they talk."

"You're absolutely right, man."

As Gonzo noted, one of the benefits of being forced to talk to every person in the neighborhood in a deliberate, methodical fashion was that it gave the people of *Abu Tayara* an opportunity to provide us with information without anyone else in the area knowing they had done so. The local people were just as concerned about the IEDs and attacks in their neighborhood as we were–maybe even more so–but they knew that if they were seen talking to us, they would be targeted for retribution and probably killed. The census ensured every person in the neighborhood got the chance to talk to us anonymously, and by doing so, provided us with a regular, reliable source of information we'd never have found any other way.

As Gonzo and I walked toward the next house on the street, the microphone on the radio I carried crackled to life, and I heard Van Awesome's voice:

"One-six, this is one-three, over."

"One-three, this is one-six, go ahead."

"One-three-alpha has found a suspected IED. Request permission to deploy the robot to interrogate the device."

"Go ahead. One-seven, you down there too?"

"Roger, I'm on site."

"We just finished an engagement, so we'll come your way and link up with you."

I walked over to SSG Fox's truck, and he led the section across 838 to a clearing three blocks north of the southern end of the neighborhood. The clearing was a fifty-yard square of undeveloped, sandy dirt that sat along the neighborhood's main street at the end of a block of houses. It was full of trash, but otherwise empty. Normally, this part of *Abu Tayara* was crowded with local foot traffic, but as was usually the case when we found an IED, everything was quiet at the moment. When our trucks pulled up, the rest of the platoon was already arranged in a perimeter around the field. The guys had gotten out of the trucks and were blocking vehicular traffic to keep cars out of the area.[3] As we pulled up, I got out and walked over to McDowell's truck to see what was up.

"Hey man, what have y'all got cooking?"

"IED, sir. See that big brick in the street?" McDowell pointed at a twelve-inch square block sitting about four inches from the side of the road in the clearing.

"Is that block not normally there?"

"Nope. It's new. As soon as we saw it, I got a bad feeling about it. I stopped the section and got out my binos, and then I saw the infrared (IR) sensor in the side of it."

3rd Squad was always the best at finding IEDs. They had found over half of all the bombs we'd come across, not counting those we'd found by hitting them. I can't put my finger on why they were so good at finding them; they just seemed to have a knack for it.

I looked through McDowell's binoculars, and sure enough, there was a one-inch, circular IR sensor embedded in the side of the block. IR sensors were a relatively new (to our area anyway) means of setting off an IED. They use the same technology used in a garage door opener to create an invisible beam of light. When something like a

---

3. The locals were usually irritated when we had to block off a road, but any time we could show them the IED we were keeping them from hitting, they stopped griping and turned around.

passing Humvee broke the beam, the bomb detonated. IR sensors could be incorporated into IEDs in several ways, but, in this case, the sensor had been attached to a block of homemade explosives, then concealed by using a can of spray foam to build a brick around it. The brick was then painted brown and rolled in dirt, creating an object which was very difficult to distinguish from any of the other sand-colored bricks in the neighborhood.

"Man, McDowell, how the hell did you see that thing?"

"Dunno, sir. Eubanks said it was new, so I got out the binos and checked it out. I guess we just got mad skills."

"I guess so. You guys got the new robot?"

"Nope, that's Blanco's guys, but they're getting it out now."

I looked down the street and saw that he was right. Private Marsh, one of Blanco's team, was getting a big black box out of the trunk of his Humvee.

"Alright, I'll hang here with you to watch the proceedings."

I climbed into the back seat of McDowell's truck and watched Marsh work. The robot was a new addition to the platoon. IED use had exploded across the theater (pun intended), and units found them–or hit them–constantly. The Army was scrambling to find ways to mitigate their effectiveness, particularly since EOD teams were always in short supply. One of the solutions was to issue scaled-down bomb robots to patrolling platoons so they could deal with IEDs themselves. The robots were relatively easy to use, and they gave platoons the ability to approach suspected devices with a robot-mounted camera, determine whether the devices were actual IEDs or not, and potentially clear them with blocks of C4. In the 1-4 CAV, each platoon had sent one soldier to a day of robot training, and Marsh was 1st Platoon's guy.

Marsh was a smart, clean-cut Midwestern kid. He was a quick learner and did his job well, so he was a natural choice to send to robot school. He'd never used the robot before today, but as an enthusiastic member of the Xbox generation, driving robots came naturally to him. In just a few minutes, the robot was out of the box and ready to roll, with Marsh controlling it from the back seat of Blanco's truck.

The robot was a smaller version of the EOD robots we'd seen in the past. It stood about two feet tall, and it had a small mechanical arm protruding from its torso and a camera for its head. I should describe it as a "he" rather than "it" because that's how we came to think of him. As the robot rolled forward toward the IED, the entire platoon watched him with bated breath. At that moment, he ceased to be an inanimate object, and became the member of 1st Platoon who had been saddled with a nasty job that nobody else wanted to do.

The robot's approach was initially confident, but he slowed his pace a bit to roll across the curb and approach the suspicious brown block that was the objective of his mission. When he arrived, he took a few minutes to visually inspect the block. His inspection revealed no wires or detonation mechanism other than the IR sensor, so he concluded that there was little he could do to disarm the device. He would need EOD support. Unfortunately for the robot, his higher command–us–was uninterested in waiting for EOD. The platoon had a good bit more work to do in sector before our shift was over for the day and waiting three or four hours for EOD would derail our schedule. When EOD arrived, they'd just blow the block in place anyway, so if we could detonate it ourselves without them, that would be better for everyone. Everyone, that is, except the robot.

Through Marsh's deft fingers, we transmitted this message to the robot and directed him to probe the device physically. The robot turned briefly to regard his platoon mates, who were watching his actions through the bulletproof screens of their Humvees. His camera eye looked almost sad for a moment, no doubt as thoughts of his thousands of robot siblings crossed his mind, then assuming a look of determination, he pivoted back to the device, raised his mechanical arm, and prodded the IR sensor with it.

The IED exploded cataclysmically, releasing a shockwave of energy that broke several nearby windows, sent a plume of smoke high into the air, and showered our vehicles with dirt and rubble. There was no sign of the robot, other than the trailing end of the wires that had once connected him to Marsh's truck but were now lying in the dust near a small crater. When the dust settled, I got out of the truck and walked over to Blanco's truck with McDowell. Chaz met us there as Blanco and Marsh stepped outside of their truck to talk to us. I turned to Chaz and said, "I guess McDowell was right. That was definitely an IED."

"Sure was. That was a big one."

"Yep. You gotta feel bad for the robot. He took one for the team."

We all nodded, laughing. He had indeed taken one for the team. As we stood there recounting the anthropomorphized version of the story we'd just witnessed, we heard shouting to the southeast. We turned to look at the source of the noise and saw a fat, middle-aged Iraqi woman yelling at Tucker. I called for Ozzie and walked over to see what she was upset about. When we arrived, she was still yelling, accentuating her words with a furiously pointed finger as she vented her anger at Staff Sergeant Tucker's impassive visage. I turned to Ozzie and said, "Ozzie, what's she so excited about?"

Ozzie and the woman spoke briefly, then turned to us and said, "She wants us to follow her to her house so she can show us something."

"Okay, lead on." I gestured to the lady to lead the way.

She did, and we followed her down the street to her house. As we waited at the gate, she walked into her yard, picked up the shattered remains of our robot, and threw them in a heap at our feet. The robot was in bad shape, but for the most part, he was still in one piece. We thought he'd been disintegrated in the blast, but instead, the IED had launched him in a hundred-foot arc that terminated in this woman's yard. The woman continued her harangue. I turned to Ozzie and waited for the translation.

"She's pissed off that we blew up our robot. She says it broke her windows, and she wants money to fix them."

"Wait, she thinks I blew up the robot on purpose? Why the hell would I do that?"

Ozzie shrugged. "She thinks we were testing some sort of weapon or something."

I resisted the urge to laugh. "Ozzie, tell her we were clearing an IED that some terrorist put on the street. That IED is what exploded, not our robot. She should be thanking us for clearing the explosive before one of her kids found it."

Ozzie translated for me. The woman stopped yelling, crossed her arms with an angry expression on her face, and continued talking.

"She doesn't believe you, LT. She just wants money."

I nodded my head. I understood. She just wanted money. "Alright, Ozzie, tell her this. That IED wasn't there when we got into sector this morning, which means it was installed in the last hour or so. Nobody else was walking around the area when we found it, which means they already knew it was there when we found it. That tells me the people in this area know who put in the IED." I paused for a moment so Ozzie could translate. The woman started to reply, but I cut her off. "If she wants money to fix her windows, tell her to tell me who emplaced the IED this morning. Even though that person damaged her house, and I didn't, I'll be happy to compensate her for the damage—but only if she tells me who did it." I stopped talking and waited for the translation and reply.

As Ozzie spoke, the woman's obvious anger was replaced with concern. She shook her head, turned, and walked into her house. When she was gone, Ozzie started laughing. "That was good, LT. Don't take any shit from her. She knows who put in the IED, but she won't tell you, so fuck her."

I shook my head. "What the fuck, Ozzie? What's up with your people?"

"I don't know, LT. This is how it is in my country. These people don't care about taking care of their own problems. They just want money, and they know you'll probably pay them if they blame you for this kind of shit." Ozzie accentuated his comment by finishing the cigarette he was smoking and flicking it into a pile of nearby trash. It was, indeed, his country. We picked up the robot together and walked back to the truck.

We patrolled for another few hours until we were replaced by 2nd platoon. During the handover, Lieutenant Smith and I spent a few minutes talking about the day. Rich was a smart guy with an amazing memory for faces and places, so unsurprisingly, when I told him about my conversation with Hamed, he told me he'd already been working a similar effort.

"Yeah, Dan, I met a guy named Dr. Manza yesterday who ran through a list of shitheads in the neighborhood with me, and I pushed them up through the S2 to see who would come up on the naughty list. He seems like a straight shooter, so we're going to talk to him again today to try to get a better understanding of the terrorist network operating in the neighborhood."

I nodded and wished Rich luck, but my attention was already wandering. After nine hours in kit, my brain was about done for the day, and I was ready to unwind. The platoon drove home without incident, and Chaz and I walked to chow with Ozzie. We ate and walked back to the barracks, where someone had already posted a memorial flyer for the robot in our platoon hallway. It seemed he'd been posthumously awarded the Distinguished Robot Service Medal. It had been presented "on behalf of a grateful mainframe" to his weeping robot wife, and his obituary had been posted in his hometown newspaper on Cybertron. I shook my head and laughed as an old quote (probably falsely) attributed to Josef Stalin popped into my head: "Dark humor is like food, not everyone gets it."

14

Route Senators, Iraq, August, 2007

"I could totally take a gorilla. He's got carotid arteries, right? That means he can be choked out like anything else. Once I've got him in the rear naked choke, he's done."

"There's no way in hell you could take a gorilla. Have you seen the arms on one of those things? He'd tear you in half."

"Bullshit. He's a dumb animal. I'm a rational human. I'd outsmart him."

"You're right, he's a dumb animal. That's part of the problem. He doesn't have rational limits on his behavior. You remember the news story about that chimp who tore the lady's face off a few years ago? That's what animals do. They tear off faces. And that was a little chimp. Gorillas are huge."

"I don't buy it. I could totally win."[1]

The enthusiastic chatter of the platoon's morning bullshit session filled my ears as our trucks drove east on Route Senators on our way into sector. Route Senators, our name for the Doura Expressway, was a divided six-lane highway that ran east-west through our squadron's area of operations, and it looked a bit different than when we arrived five months earlier. The T-wall project the support company had been working on was complete now, so the entire length of Senators was lined by an unbroken line of twelve-foot-tall concrete barriers. Like concrete walls in urban areas everywhere, the T-walls had already accumulated a thick coating of stylized Arabic graffiti, most of which was probably anti-American. It wasn't aesthetically pleasing, but it kept down

---

1. This long-standing, man versus animal debate was eventually resolved. According to the internet, an adult male gorilla can bench press around 4,000 lbs. The strongest guy in the platoon couldn't lift more than 8% of that. Once this overwhelming strength disparity was established, the "I could take him" side of the argument admitted defeat.

the number of IED attacks on our convoys.

Today's assignment was to finish up the last few streets of our census work in 838. I was back in Tucker's truck today, and as it roared along the highway, I reviewed the paperwork on the houses we were supposed to visit. Private Smith, our platoon commo guy, was in the seat next to me, telling me about his desire to open a barbershop after redeploying.

"I spent a lot of time in barbershops growing up, sir, and I really like the culture and feel of them. I want my shop to be a place where folks can come and enjoy themselves while they wait for their haircuts."

"Sounds fun, man. I guess for me, getting a haircut has always been a pretty mundane experience."

"A good barbershop has a great feel to it, sir." He paused and looked up. We were approaching the elevated clover-leaf intersection at the southern end of *Abu Tayara,* and there was a Stryker convoy on the overpass above Senators. "I guess the Stryker guys are headed into sector today too."

Strykers are eight-wheeled armored personnel carriers that were still a new sight in 2007. There were several variants, but the Strykers crossing the bridge in front of us were long, low, turretless vehicles with a machine gunner poking out of the front hatch, and a troop compartment with nine infantry guys in the back half. Originally designed to be a lightly armored vehicle, these Strykers had been outfitted with aftermarket armor packages to protect them from IEDs. Compared to our Humvees, which could hold only two dismounted soldiers each, Strykers were great for the counterinsurgency fight because they could carry a large number of soldiers in a relatively small number of transport vehicles. This allowed Stryker units to search more houses and engage with more people since fewer people had to crew the vehicles.

We were about 200 yards from the overpass when an enormous explosion suddenly rocked our convoy. The shockwave made my vision go blurry for a second, and as my eyes cleared, I watched a Stryker fly sideways off the overpass. The second of the four Strykers crossing the bridge had hit an IED embedded in the median, and the blast had launched the vehicle through the metal guardrail and off the bridge. As it was falling, I experienced one of those slow-motion moments you see in movies where I watched the machine gunner fly out of his hatch and fall the twenty-five feet to the street below alongside the plummeting Stryker. The Stryker smashed into the street, releasing a huge plume of dust that kept me from seeing the gunner hit the ground.

"Shit," I thought, "That dude is dead. We're going to need a shovel to dig him out." I picked up the radio and reported the incident to higher as our platoon raced over to assist the stricken vehicle. The remainder of the Stryker platoon had to drive the entire length of the cloverleaf to get to their damaged vehicle, so we beat them to the site. While our gunners secured the fallen Stryker, the rest of the platoon dismounted to start helping the casualties.

"Grab a shovel!" I shouted at Smith as we jumped out of the truck to run over to the wreck. He started pulling the tools out of the trunk, and Chaz and I ran over to the soldiers who were slowly crawling out of the upside down Styker. I saw one guy lying on his back in the dirt and thought he must be the gunner I had watched fall off the overpass. Incredibly, as I walked over to him, he stood up and dusted himself off.

"Hey man, are you okay?"

"What? Yeah. I think so. What happened?"

"You fell off the overpass, man. You guys hit an IED, and you fell off that bridge." I pointed up at the bridge. He turned his head to follow my finger, looking up at the overpass, and then looking back down at me.

"Really?"

"Yeah, man, really. Here, come sit in our truck. We need to get you to the hospital."

Blanco had walked up while we were talking, and he took the dazed soldier to one of our trucks and helped him sit down. I found the lieutenant from the Stryker unit who'd just arrived with the rest of his platoon and told him we'd MEDEVAC his guys to the CSH for him. He agreed, and we loaded up the five guys who had been in the fallen Stryker and drove north to the hospital while the Stryker platoon took over security and started to coordinate with their higher headquarters for a wrecker to recover the destroyed vehicle.

The drive to the main hospital in the Green Zone was short, and in fifteen minutes, we arrived at the emergency entrance to the CSH. Tucker and I stood outside the truck watching as the medical team swarmed our vehicles to pull out the injured guys, strap them down to gurneys, and rush them into the hospital for treatment. The fallen gunner looked a little confused, but otherwise unhurt. I waved at him as the medical team whisked him away, but he didn't seem to notice.

"Hey Tuck, did I imagine that guy falling off the overpass?"

"Nope. That guy fell thirty feet, easy. I thought he was going to be a puddle when we found him."

"Me too, man. I thought for sure he was a goner."

"That's some crazy shit."

It was indeed crazy. How had that guy survived? I couldn't stop turning the question over in my head. I saw some crazy things during my time in Iraq, but none of it was crazier than watching that soldier get blasted off an overpass, fall twenty-five feet to the ground, and then stand up and walk away from the incident.

The drive back from the CSH took us by the famous Iraqi Victory Arch–officially named the *Swords of Qādisīyah*–that Saddam Hussein had built to commemorate the 1980-1988 Iraq/Iran war. Our platoon had stopped there in July to reenlist Staff Sergeant Fox and take platoon pictures under the huge pair of crossed sabers, and while there, we had broken into the old reviewer's stand overlooking the monument so Fox could reaffirm his allegiance to the United States at the very same podium Saddam Hussein once used to preside over ceremonies while firing his shotgun into the air. It had been a good day.

Today, though, we sped past the monument to get back into sector. The IED had derailed the day's plans, and we needed to get back to work. The drive south was uneventful, and in a few minutes we were weaving back through the serpentine pattern of T-walls which marked the entrance to *Mahala* 838. As always, the Iraqi police who manned the checkpoint waved us through. As we passed, I gave the police chief a thumbs up, and he returned my gesture with a smile and a wave. Our relationship had improved substantially since our days in Mechanics. The police appreciated our willingness to patrol the dangerous neighborhood, and we valued the checkpoints they maintained at the two entrances to the *mahala* because they discouraged (but didn't prevent) entry by the enemy.

After clearing the checkpoint, the platoon split up, and Tuck, Blanco, and I headed south to get to work on the census. About halfway down the main road though, an old woman stepped away from the crowded foot traffic on the sidewalk and flagged us down. It looked like our assignment would have to wait. We'd met the old woman on a previous engagement, and Ozzie remembered her well, so we stopped the trucks and got out to ask her what she needed. As we walked up, she immediately launched into an excited string of Arabic. Ozzie translated.

"LT, she says someone put a bomb in a house on her street. She wants us to take care of it."

I nodded, said "Okay," and asked her to show us where it was. She started weaving her way through the crowded road toward a street to the southeast. Ozzie and I followed on foot, and Tucker's section maneuvered its vehicles behind us. House-borne explosive devices, or HBIEDs, were a fad in the Baghdad terrorist community this month, and they were a real problem. Baghdad was full of abandoned buildings, and seeing U.S. forces regularly search them, terrorists recognized an opportunity. By filling a building with a huge pile of explosives, and then rigging a detonator to one or more of the entrances, a well-built HBIED could collapse the entire structure as forces entered the building and potentially kill them all.[2]

As I followed the lady to her street, I discussed potential solutions to the problem on the platoon net. Several ideas were passed around, to include climbing in through a second-story window or calling for an EOD team with X-Ray equipment, but the best recommendation came from Van Awesome.

"Don't go in the house at all, sir. If there's a bomb in it, it'll probably kill the team trying to find it. Just throw locks on the doors and post a 'keep out' sign so people stay out of it."

"What if someone breaks in?"

"Then they explode. If they're breaking into houses, they're criminals anyway, and it's better them than us."

Van's idea made sense. Our chances of successfully finding a bomb in the house without setting it off were low. Even EOD techs would struggle to do so, and we didn't have their tools or training. Maybe the best answer really was just to board the place up and avoid it entirely.

Unfortunately, as we made our way southwest, the point became moot. An enormous explosion rocked the neighborhood, and falling bricks began raining down around us. The woman began howling and broke into a run; the explosion had come from her street. We ran after her, and as we turned the last corner, we saw that her report on the HBIED had been accurate. One of the houses on the southern side of the road had

---

2. This exact situation happened to my brother-in-law, a sniper assigned to a Strkyer unit which was also operating in Baghdad. He was on leave in the United States when his section encountered an HBIED in a sector south of us. The snipers had broken into a building to set up an overwatch position for a clearance operation, but while doing so, they had set off a boobytrap rigged to the front door. The bomb blew up the entire building and killed the entire section.

exploded, leaving a smoking concrete frame where a two-story house had stood a few moments before. I got on the platoon net.

"I need Doc down here, ASAP."

Dan "Doc" Osborne was our platoon medic. He was a Southern guy with glasses, a thick accent, and a dry sense of humor. He was also an exceptional medic, and, based on the size of the blast I'd just seen, we were probably going to need his talents. He and Chaz were with 3rd Squad for the day, and they pulled up to the scene just as I arrived on foot with Ozzie and the old woman.

Our group assembled on foot in front of the remains of the building while the trucks secured either end of the road and handled crowd control. The woman shouted something none of us could understand and waved her arms a bit, then walked back up the street to her house–presumably to check it for damage. The remains of the house in front of us were a smoking ruin. Like most of the houses, it had been a two-story house with heavy brick and concrete walls. Most of that concrete was now lying in a huge, crumbled pile that dribbled out into the street in front of the shell of what had been the house. Bits of furniture and rebar protruded from the debris at odd angles, and shards of broken glass were scattered over the whole mass like the sprinkles on a sundae.

The strangest part of the scene, though, was the silence. I imagined we would arrive to find a crowd of distraught onlookers, but once the old woman who was with us left, there were no locals present. Other than the distant rumble of our Humvees' diesel engines, there was no sound at all. That silence was broken when Doc spotted something.

"Hey, sir, there's a foot." He gestured with his weapon to a spot on what used to be the first floor. We could see the foot and leg–clad in brown pants and a flip-flop–but the rest of the body was buried under the rubble.

"Let's clear him off and make sure he's not alive."

Doc looked at me with one eyebrow up. "Sir, he's dead."

"Nope. We don't know that yet. Let's check."

While Chaz and 3rd Squad checked the rest of the rubble, Doc and I walked over and cleared a large piece of concrete off the dead body. The rest of him was burned to an unrecognizable crisp.

"Check his pulse, Doc."

Doc looked at me and shrugged as he shook his head and smiled. He reached a gloved hand down to the body's blackened neck and put two fingers on it. Then he turned to me and said, "Yep. He's definitely dead."

I smiled and replied, "Now I believe you, Doc. Let's get the police to clean this guy up."

The first time we'd found a body together, even though Doc had asserted he was dead with only a casual glance, I'd thought the guy might still be alive and told Doc to confirm his diagnosis manually. He'd done so, and sure enough, the guy had been dead. Doc had a good eye. Since then, we'd reenacted the same scene dozens of times, even when the bodies were so far gone that their continued life would have been fodder for a zombie movie. This reenactment was a thing Doc and I did. The ritual brought a bit of levity to an otherwise horrific aspect of our job, and it made the situation a bit more manageable.

"Chaz, can you grab the police and clean this guy up? I need to get going with our census mission." The Iraqi police still didn't have great relations with the local people, but they were good at this sort of cleanup work, and by letting them do it, we gave them the chance to show their higher headquarters that they were staying busy and being effective.

"No problem, sir. I gotta swing by the COP and poop real quick though." A long and colorful Army career had left Chaz with notoriously irritable bowels.

I nodded and called Tucker to come pick me up. There wasn't anything else to do with the HBIED but clean up the mess it had left, and continuing to stare at it wouldn't accomplish anything.

We found out later that Van Awesome's hypothesis had been correct. The dead guy was a vagrant who had detonated the HBIED while breaking into the house. My thought on hearing that news was, "Better him than us." Thinking back about that now, I feel a bit bad about my perspective. That HBIED had been emplaced to kill us, not vagrants. We were there to protect people and restore stability, and I wish we could have prevented more of the locals from being killed in the crossfire. Still, Van Awesome's reasoning was sound. If we had investigated and reduced all the HBIEDs we found that year, we would probably have gotten soldiers killed, and I'd rather let a vagrant die because of the enemy's actions than get one of my guys killed because of my own.

1st Squad pulled up next to me, and as I climbed into Tucker's truck, I pulled out the sheaf of paperwork from headquarters on our census assignment. I had just confirmed the street we were supposed to survey for the third time that day when another explosion shook the neighborhood.

"Now what?" I thought. I threw the truck's headset on my head and slammed the truck's door shut. The platoon net was already buzzing with reports about the explosion. 2nd Squad reported they'd seen smoke and heard an explosion and rifle fire from the southern police checkpoint; someone had attacked it.

As our truck sped down the street, I keyed the microphone and said, "All Comanche-one elements, let's get down to the southern checkpoint and see what happened." The platoon was way ahead of me, and as 1st Squad and I weaved through the checkpoint's barricades, I saw the rest of the platoon's vehicles in a perimeter around an Iraqi police vehicle that was burning in the middle of Route Senators, only a few hundred yards from where the Stryker had landed earlier that morning. The tall, armored SUV had been parked in the middle of Senators scanning the ruined buildings south of the checkpoint for enemy activity when it was hit by an RPG fired from Mechanics. The truck was now a roaring inferno. The police who manned the checkpoint were huddled behind a concrete barrier. They wanted to get to the truck and pull out their injured comrades, but they were terrified to approach it. We needed to give them a hand.

I directed Tuck to pull up next to the burning truck, and I keyed the microphone again. "One-two and One-three, suppress any enemy you see while One-one and I help the police get their guys out of this truck. One-seven, we'll probably need Doc."

"One-two, roger."

"One-three, roger. We'll push up to the overpass to get better observation."

"This is One-seven, roger that."

I acknowledged their transmissions, then got out of the back of Tuck's truck. I was about ten feet from the burning police truck with our armored Humvee between me and the neighborhood to the south, where the RPG had originated. The fire's heat was palpable, and machine gun fire ripped through the air as 2nd and 3rd Squads' guns "talked" to one another during their suppression mission. I quickly took in the situation.

The rear door of the armored SUV was closed, but if there was anyone alive in the truck, the rear compartment was where they'd be. I looked back at the police, who were still huddled behind the checkpoint's barricades about thirty feet north of me. I yelled across the road at them and waved my arm vigorously. "Get your asses up here and help me out!" They didn't speak English, but my meaning was clear, and while our trucks continued suppressing Mechanics, four Iraqi policemen grabbed a stretcher and ran out to join me at the burning truck. Doc had arrived by then too, and together with the Iraqis, we tore open the truck's back door.

Smoke poured out of the back, and three or four rifle rounds that were cooking off from the fire's heat zinged past us. Through the smoke, we could see two Iraqis unconscious on the floor.

"Grab them!" I shouted at the stretcher team. The Iraqis ran forward with the stretcher, and one of them, shielding his face from the heat, pulled one of the unconscious guys out and laid him on the stretcher. Doc and I grabbed the other guy by his arms and legs, and the six of us ran back toward the checkpoint. When we arrived, the rest of the police quickly gathered around their fallen comrades, several of them carrying the casualties into the shaded headquarters area while the rest cried and wrung their hands. The Iraqi lieutenant approached me and grabbed my hand. He said something in Arabic I couldn't understand, but he was clearly grateful for our help. I nodded and turned to Doc.

"Doc, go help those guys out."

Doc turned to me and gave me the same bemused look he'd given me earlier.

"Sir, those guys were dead as hell. Did you see how burned they were?"

I stood there for a minute, thinking.

"No, I didn't notice. Were they bad?"

"Sir, I've seen barbeque that was less well-done than those Iraqis were."

"Did you check their pulse?"

Doc smiled. "No, do you want me to?"

"Nah, I believe you this time. Go in there and make it look like you're trying to help them out though. I think they'll appreciate the gesture.

Doc nodded and walked into the checkpoint. As I stood there thinking, Ozzie and Chaz walked up.

"Hey Chaz, can you and Ozzie go with Doc and see if he needs any help with the police?"

Chaz nodded. "No problem, sir. We'll take care of it."

As they walked toward the checkpoint, I checked my watch. It was almost time for 2nd Platoon to relieve us. "What the hell?" I thought. "Where did the day go?" We had been in sector for more than eight hours, and we hadn't gotten a single thing done from our original assignment. I walked back to Tucker's truck and climbed in to wait for Ozzie. The gunfire from Mechanics had stopped, and while I was dealing with the casualties, Van Awesome had reported the incident up to headquarters so the unit to our south could search the area where the gunmen had been.

Doc and Chaz finished up at the checkpoint and walked back over to my truck. Chaz spoke up.

"Hey, sir, those two Iraqis were dead as hell, but the police are super happy with us for helping out. There's not much more we can do for them. We should probably get ready for the handover with 2nd."

"You're right, Chaz. Man, what a day this was."

Chaz nodded. "Yep. Pretty fucked up for sure. I think it's stir-fry night, but I might skip it after seeing those two dudes."

We climbed back into our trucks and pulled back through the checkpoint onto Main Street. 2nd Platoon arrived, and I caught Rich Smith up on the day's activities. I handed him my neglected sheaf of papers from headquarters.

"Rich, I didn't get shit done today on the census. You want to give it a shot?"

"Sure thing, brother. I'll see what we can get done."

We shook hands, and I headed back to the truck and told the platoon to head back to base. The drive back to Falcon was short, and blessedly uneventful. When we got back, I finished my closeout reports and walked back to my room to drop my sweat-soaked gear. I stripped down to shorts and flip-flops and checked my watch. It was morning back home, and I could probably catch Alycia if I called now.

I rubbed my eyes and thought about the call. On the one hand, I wanted to call and see how Alycia was doing. On the other hand, I worried that if I called, we'd argue about something. It was difficult to avoid friction. Alycia would tell me about trouble with Gareth, or a call she had to make to the phone company about a billing issue, and I'd listen and try to be supportive, but the problems often seemed trite in comparison to what I was dealing with in Iraq, and it was hard not to be mentally pulled away by my own concerns. At the same time, I was restricted in what I could talk about because discussing operational details was prohibited, so my side of our conversations was always vague and guarded. Aside from the prohibition, I didn't even *want* to talk about a lot of the stuff. Why would Alycia want to hear that I'd watched some kid fall off an overpass? Would she be able to relate to the inside joke I had with Doc about checking the pulse on dead bodies we found? *Of course* she wouldn't be able to. Nobody back home could, and nobody wanted to hear about that stuff. That's what my days consisted of, though, and it was hard to think about anything else.

After some thought, I decided to skip the call. It'd be better not to talk at all than to have an argument. It had been a long day, and I needed a shower and food. We still had another ten months until I'd be home. What were our conversations going to consist of by then?

15

Mahala 838, Iraq, September, 2007

As much as I hated to admit it, it looked like area beautification might save our lives.

In terms of U.S. casualties, 2007 was the most dangerous year of the war in Iraq, and that summer Baghdad was one of the deadliest places to be.[1] Like the rest of the 1-4 CAV, 1st Platoon had been working ten-to-fourteen-hour days, every day, to meet the people in *Abu Tayara,* arrest the terrorists hiding there, and clear the IEDs the bad guys had emplaced in the neighborhood. Through a combination of good luck and solid work, 1st Platoon had suffered no fatalities, but the stress of this long, hot summer had taken its toll, and we'd rotated out a few personnel. Fox and Tucker had hit IEDs and needed to take a knee, so we moved them to the Company headquarters to help run the administrative and intelligence cells there. Chaz's unorthodox personality and love of glorious hair styles had finally run so far afoul of the Squadron and Brigade Sergeants Major that even my direct appeal to Lieutenant Colonel Crider couldn't save him, so he was moved up to the squadron operations shop as penance for his dislike of hair and uniform regulations and his overdeveloped sense of humor. Our new platoon sergeant–2nd Platoon's freshly promoted Sergeant First Class Kaluzny (of earlier roof-top argument fame)–gave 2nd Squad to Staff Sergeant Gonzales and 1st Squad to Sergeant Blanco.[2]

With these changes, the platoon continued its efforts through August and early September. Our initial survey of the neighborhood was complete, which gave us the

---

1. *globalsecurity.org.* 2007. https://www.globalsecurity.org/military/ops/iraq_casualties.htm.

2. Kaluzny and I quickly overcame our initial friction from COP Amanche and ended up getting along well. We still do to this day.

sense that we were making progress. But the neighborhood was still a violent place. In addition to the IEDs we were still finding daily, our opponents had added a new weapon to their arsenal: the RKG-3 grenade.

When the RKG-3 exploded, it produced an armor penetrating shaped charge similar to the EFPs we'd been encountering for months. Instead of being a victim-activated booby trap, though, the EFP grenade was hurled directly at its intended target, usually from behind a wall so the thrower couldn't be seen. A small parachute on the rear of the grenade ensured that the grenade was oriented the right way when it hit, and the resulting explosion instantly cut through any vehicle's armor. The grenades had already taken their toll. In the last four weeks, Apache Troop had lost two soldiers to them, Staff Sergeant Courtney "Tuba" Hollinsworth, and my old armorer from back at Riley, Specialist Long. Several other old friends, including Banninger and Phillipus had been injured and evacuated from theater. So far, Comanche Troop had been lucky. But the recent deaths of guys I had been close to in Apache Troop hit me hard, and I felt an enormous amount of pressure to stop the attacks before our platoon suffered fatalities, too.

Since the grenades were being thrown from behind the courtyard walls, and because the courtyards chosen were always those around abandoned houses, the main preventive thing we could do was to reduce the number of abandoned houses in the neighborhood, best accomplished by encouraging friendly families to move into them. Similarly, since trash and rubble provided excellent places to hide IEDs, anything we could do to clean up the neighborhood would reduce the enemy's ability to emplace bombs. Ironically, although 'area beautification' was one of the most loathed and useless-seeming missions a military unit could be given, making *Abu Tayara* beautiful was critical to keeping us alive.

To an extent, the census was a critical first step in this process. It showed us where the abandoned houses were, and it introduced us to the local businessmen in the neighborhood who might be interested in getting to work on the problems we identified. Unfortunately, understanding the problems wasn't enough. *Mahala* 838 was still as dangerous for the Iraqi people as it was for us. Shi'a terrorists were still issuing death threats to the Sunni families living there, which encouraged them to leave and kept former residents from moving back into their homes. For similar reasons, the local business owners were unwilling to reopen their businesses, and even if they had been willing to do so, they had no money to proceed. As we continued to play whack-a-mole with terrorists and IEDs in *Abu Tayara,* we realized that if we wanted to achieve any lasting effects, we would need to do something else.

As we drove back from the Detainee Holding Area, or DHA, that September morning, that something seemed a long way off. We'd been at the DHA dropping off a group of detainees we'd arrested that morning. We'd gotten the names and addresses of the bad guys from a family we'd met during a census engagement, and the raid we'd conducted to pick the bad guys up had been a 'cordon and knock' operation where we'd pulled up around the house, knocked on the door, and arrested the identified people. Over the course of the deployment, we'd figured out that if we knew where the bad guy's family lived, it was safer and more effective to simply knock on the door to his house in the early morning while he was home and asleep than it was to conduct a nighttime raid. In our experience, the bad guys didn't want to put their families at risk any more than we did, and night raids put everyone on edge and increased the odds the wrong people would get killed.

The drive back to *Abu Tayara* took around twenty minutes and passed without incident. As we drove back through the police checkpoint, Dr. Manza stepped into the street and waved at my truck. Dr. Manza was an Iraqi cardiologist whom Lieutenant Smith had first bumped into a few weeks earlier. A nondescript man in his late thirties Dr. Manza was of medium height and build, with a moustache that would make Tom Selleck jealous, and an unassuming, friendly manner which made him easy to talk to. We'd talked to him a few times during our population engagements, but I didn't have any special relationship with him. Still, he clearly wanted something, so we pulled over, and Gonzo and I stepped out to see what he needed. Dr. Manza spoke fluent English, so Ozzie didn't come with us.

"Hey Doc, what can we do for you today?"

"Good morning, Lieutenant Dan. There is a *qunbula* in a pile of trash near my street, and I would like you to destroy it please."[3]

Hearing that sentence in English on this sunny morning was bizarre. It took me a second to digest it. "Sure, Doc. We can do that. Can you show it to me?"

"Yes. Please follow me. Also, I know who put the *qunbula* in the trash last night. Would you please arrest them too?"

"Sure, Doc. No problem."

---

3. *qunbula,* pronounced "coon'-buh-lah" is bomb in Arabic.

Dr. Manza started walking north toward his street, and I followed him, calling 2nd Squad on the radio to tell them to send a few dismounts with me. Usually, Iraqis were very touchy about providing us with the names of bad guys because they feared reprisals, so I assumed Dr. Manza would want to go somewhere private to give us the information. We walked a few blocks through the busy morning foot traffic in silence until we reached a small shop on the west side of the road that sold snacks, drinks, and pirated DVDs.[4] We entered the shop together, and Dr. Manza pointed at the two smiling young shop keepers and said, "Those are the two men who put in the bomb. Arrest them please."

The two men didn't speak English, so they continued smiling. Gonzo and Schryver looked at me with artificially neutral expressions on their faces. I stood blinking at Dr. Manza for a second, processing the situation. It had developed in a way I hadn't expected and had become incredibly socially awkward.

I nodded before asking, "Do you have anyone else who can confirm this information?"

Dr. Manza nodded and said, "Yes. I can get statements from three other people on my street who also saw them do this."

This was all a bit unorthodox. Normally, we needed at least two people to sign sworn statements that someone was a criminal before we'd arrest them. This reduced the potential we'd be used by the locals to settle personal vendettas. Technically, I should have left the shop and had Dr. Manza collect the statements first. Then, I should have scrubbed the statements to ensure he wasn't either completing them all himself or having his friends and family members corroborate his story. Given the circumstances though, none of that was practical. If we left this room without arresting the men, they would almost certainly flee the area. Moreover, Dr. Manza had put himself on the line by giving us this information. If we left now, he would almost certainly never do so again. We needed members of the community who were willing to help us solve the security problem in *Abu Tayara*, and I felt that turning my back on Dr. Manza today would be a mistake. I figured, worst case, the two men were innocent, and if so, they'd only spend a few days in U.S. custody before being released.

I nodded and said, "Okay, Gonzo, please arrest these two guys and get them ready to

---

4. The pirated DVD market in Iraq was extensive, and you could find everything from '80s movies to new releases for only a few dollars apiece. The labeling could be unreliable though. I once bought a copy of *The Seven Voyages of Sinbad* that turned out to be Season 2 of *Sex in the City*. COP life being what it was, I watched it anyway and ended up enjoying it enough to buy the other seasons from the same shop.

take to the DHA."

Gonzo called his guys on the radio, and the two young men's smiles quickly faded as Gonzo and his team bound their hands behind their backs with flex-cuffs and escorted them out of the shop to be stuffed into the back of two of our vehicles. The men didn't resist or shout, and as they were led into the busy street, nobody seemed particularly taken aback. I took this to be a good sign and turned to Dr. Manza and handed him the blank forms I needed him to complete for the detainees.

"Okay, Doc. Can you show me where the IED is? We'll take care of it while you get the paperwork ready on these guys. I need you to fill out one of these forms for each guy, and I need you to collect written statements from the other people who saw them put the bombs in the trash piles."

Dr. Manza took the papers, folded them in half, and put them in his jacket pocket. He seemed remarkably at ease with the situation, given that he had just turned in two local boys to an occupying foreign army and was standing in the middle of a crowded street in his neighborhood. "Okay, Lieutenant Dan. I will show you where the *qunbula* is now."

I gave him the universal "after you" hand gesture, and we walked west along a side street for about a block to a vacant lot between two houses. Since 2nd Squad was busy with the detainees, 3rd Squad's vehicles followed us to provide any necessary support. The lot was an empty dirt field framed by the courtyard walls of the adjacent homes, and it was covered in a 12-inch-thick layer of onion paper, food packaging, and other miscellaneous garbage. It was a perfect place to hide an IED. When we were about fifteen yards away, Dr. Manza stopped and pointed at the field. "It's in there."

"Okay. Did you see what the IED looked like by chance?"

Dr. Manza shook his head. "No. I just saw them slow down the car and throw something in the trash before speeding off."

"You're sure it was a bomb, though?"

"Yes. Very sure."

"Okay. We'll figure it out then. Thanks, Doc. We'll come by your house once we're done to grab that paperwork."

"Okay, Lieutenant Dan. I'll see you then."

Dr. Manza walked back down the way we had come and disappeared from my view behind 3rd Squad's trucks. McDowell's truck was the lead vehicle, so I walked over to his door and met him and Van Awesome to talk through how we'd find and clear the

IED.

"Alright, guys, let me see what our EOD response time looks like today." I keyed the mike and said, "Comanche seven, this is six."

Kaluzny's voice came over the net. "This is seven."

"Mind calling EOD and getting a read on how long it'll take them to respond?"

"I just did. Bad news. Looks like we're probably looking at more than three hours. Want me to request it?"

I turned to Van Awesome and McDowell. "Fuck. I don't want to sit here all afternoon waiting on EOD. What do y'all think?"

McDowell grinned broadly. "Let me try out the new mine roller, sir."

The mine roller was McDowell's latest toy and one of the newest tools the squadron had received to combat IEDs in sector. Because of the damage IEDs inflicted on U.S. forces, the IED countermeasure effort was a major Army undertaking. In addition to the robots we'd been issued, our platoon had received bomb-proof suits, bomb-resistant underwear, and shark fin-like contraptions that mounted to the front of our trucks to disrupt EFP detonation.5 We'd covered our turrets with camouflage netting to deflect grenades and strapped cans of water to our doors to mitigate shaped charge damage. Most of these measures were effective in their way, but the enemy was clever enough to adapt, so our countermeasures always had an expiration date. Since so many of the recent IEDs we'd found were activated by pressure-plates, and because McDowell usually took point on our convoys, he had decided to draw a Self-Propelled Adaptive Roller Kit (a.k.a. 'the mine roller') from the motor pool. The mine roller was 6,500 lbs. of steel and rubber rollers that attached to the front of the Humvee and made it look like a giant lawn mower. It was designed to roll over everything in front of the vehicle to set off any mines—or IEDs—it encountered.

I looked at the field of garbage and envisioned what McDowell had in mind. The roller was designed to be used defensively to keep the convoy safe from anything we would incidentally hit while driving our normal route, but this looked like an opportunity to use it a bit more aggressively and to save ourselves time in the process.

"Are you thinking you'll just roll back and forth through that garbage until the roller sets off the IED?"

---

5. The bomb suits came with giant, water-bug-like overshoes that were notoriously difficult to walk in. The only thing we ever used them for was to run races in them on the uneven gravel surface of the FOB for comedic relief.

"Yeah, pretty much. Me and Eubanks will blow that shit up in a hurry, then we can get out of here on time."

"What about the mine roller?"

McDowell shrugged. "Nobody else uses them, so there's a pile of extra rollers in the motor pool. If this one gets fucked up, we'll just pick up another one. The roller puts so much strain on the engine that we've got to drop our truck off for maintenance tonight, anyway."

I nodded and thought through his answer. Normally, voluntarily hitting an IED wasn't an approved method of clearing it, but the roller was designed to absorb IED blasts, and the truck was armored enough to prevent any shrapnel from hitting the crew, so it didn't seem too risky. It certainly would save us a good bit of time.

"Alright, man. Do it. Just keep it slow, so if the IED has a delayed fuse, you won't accidentally put the body of the truck over the IED before it explodes."

McDowell grinned again. "Roger that, sir. We'll knock that shit out."

McDowell climbed back into his truck, and Van Awesome and I climbed into his.

"What do you think, Van? Was that a dumb call?"

From the back seat, all I could see was the back of his head as he replied. "I'm not sure, sir. I don't think there's much risk of anyone getting killed, but it's possible. I wouldn't have played it that way, though. I think it would have been better to wait for EOD."

Van Awesome was a meaty rugby player who had studied for a few years at West Point before deciding he'd rather serve as an NCO. He had a sharp mind, a good grasp of tactics, and a deadpan way of delivering pointed feedback that frequently made me feel like he probably had a better idea about how we should be doing things than I did. Van Awesome never said anything that suggested he thought ill of my decisions, but the sense that he might was frequently in my head anyway.

As I watched McDowell's truck roll toward the trash pile, the fear that I'd made the wrong decision started gnawing at me. What if I was wrong, and the IED killed someone? What if the IED was huge and took down one of the adjacent buildings? What if there was an investigation into what happened with the mine plow, and I got fired? When life or death decisions came up daily, second guessing yourself and being second guessed by other people was one of the inevitable but unpleasant aspects of being in charge. Even when a decision went well, it was hard to avoid thinking it could have gone better, and when something went poorly, everyone (including you) tended to sit around speculating about what you could have done better. Over time, this eroded confidence and, combined with the stress of seeing the often-catastrophic consequences of bad decisions, led some guys to eventually collapse.

The massive blast of an IED detonating suddenly interrupted my wool gathering and rocked our truck backward. As the suspension compensated and we bounced lightly in our seats, McDowell's cackling laughter broke across the radio.

"Comanche six, this is one-three-alpha. The lawn has been mowed. IED detonation complete. We're all good here."

"Roger that. Pull back, and let's check the plow."

As I unkeyed the radio, I felt a sense of relief as my jaw relaxed. I must have been clenching it without noticing. Van Awesome and I climbed out of the truck and walked down the street to where McDowell and Eubanks were checking out how much damage the mine plow had suffered. One set of the plow's massive steel wheels had been blown off entirely, and the hydraulic shock that had served as that wheel's suspension was leaking fluid in the street. Otherwise, the truck seemed fine.

"You guys okay? Is this thing still drivable?"

Eubanks, tethered to the driver's seat by the hose that circulated cold air through his vest, replied. "She's good, sir. We'll have to take it a bit slow on the way home, but nothing's dragging. We should be fine to get back to the motor pool this afternoon."

"Sounds good, Eubanks. I need to talk to Dr. Manza for a bit, so you guys can keep your truck parked outside his house to avoid moving it too much until we RIP. Van, I need a guy in with me. Otherwise, I need you guys to lock down the street."

"Roger, sir. Sproul can come in with you. He needs a chance to get out of the truck for a bit. The rest of us will stay outside."

I nodded, and we started walking back to the trucks to drive over to Dr. Manza's house for the meeting. On the way, we were interrupted by a ten-year-old kid in a light blue *dishdasha* and a white *kuma* who ran up and started shouting "Mistah, Mistah!" at me.[6] He gestured enthusiastically for us to follow him. We didn't have Ozzie with us, so I shrugged and motioned for him to lead on, calling Sproul, Van Awesome, and McDowell to follow me. The kid led us thirty yards to the end of the street, where he opened a decorative metal gate and ran ahead of us up a set of exterior stairs that led to the roof of a two-story house. We made our way up the stairs cautiously, in case he was leading us into trouble, but when we rounded the final corner, we found him smiling

---

6. A *kuma* is a pillbox shaped hat popular in parts of the Middle East.

down at a large green object at his feet. It was the wheel from our mine plow.

"Mistah! *Qunbula!* Candy! Candy!"

The IED's blast had somehow launched this seventy-pound wheel fifty yards down the street and twenty feet up in the air, and now this kid had found it and wanted a finder's fee for reporting it to us. I smiled at him, pulled a roll of *dinar* from my pocket, and peeled off a bill for him as Sproul and Van wrestled the wheel down the stairs.

"*Shukran Jazilan, habibi.*"[7]

The kid's smile was broad and genuine as he ran off to lord his newfound riches over the other neighborhood kids who had congregated outside the metal gate to the house. Usually, the kids would mob us to beg for candy as we walked back to the vehicles, but today they didn't even notice us. We watched them disappear down the street, chasing their newly wealthy *compadre* as he waved the money over his head and ran off to wherever it was Iraqi kids went.

We climbed back into our vehicles and rolled back to Dr. Manza's house. On the way, Kaluzny called on the radio to tell me he would leave sector with Gonzo's squad to take the two detainees to the DHA once they had Dr. Manza's statements. 3rd Squad and I arrived at Dr. Manza's house, where Sproul and I were greeted at the gate by Mr. Fasil, Doc's smiling, elderly manservant. Mr. Fasil showed us into the living room, which was a large, dark corner room with a low ceiling, marble floors, and long, red, plush couches arranged around a rectangular mosaic coffee table. A chandelier hung over the coffee table, and a bowl of Jordanian almonds sat in the middle of it. As Sproul and I walked across the room and sat down heavily on one of the couches, Mr. Fasil turned on a small air conditioner that sat in the corner and gestured for us to take a handful of almonds. He then disappeared around the corner to make each of us a cup of tea.

The transition from the world outside, with its bombs, death, chaos, and hundred-plus degree heat to this dark, cool, quiet room where a nice old man brought us snacks was jarring, but pleasant. The relative luxury of these meetings was seductive, and I had to exert mental effort to avoid prolonging them just for the sake of sitting somewhere comfortable for a while.

Sproul and I sat quietly munching on the almonds for a few minutes, before transitioning to the small *istkan* of heavily sweetened tea that Mr. Fasil brought in on a large silver tray. The air conditioning had just begun to cool our sweat-soaked uniforms

---

7. "Thanks buddy" in Arabic.

enough to produce a shiver when Dr. Manza walked in with a sheaf of papers in his hands.

"Hello Lieutenant Dan. I have the papers for you."

"Thanks Doc. Sproul, can you run these out to Kaluzny and Gonzo real quick?"

As Sproul did so, Dr. Manza picked up a cup of tea, lit a cigarette, and sat down across from me on the couch.

"I heard the *qunbula* explode. Is everything okay?"

"Yep, everything's just fine. The *qunbula* was just where you said it would be, and we cleared it without anyone getting hurt."

Dr. Manza smiled at the news. "That is good. Our neighborhood used to be very nice, but there are bad people here who have made it not so nice lately."

I nodded. "From the looks of it, this used to be a wealthy area. I can imagine it was pretty posh a few years ago. With a little work, I think we can get it back there again."

Dr. Manza's smile didn't fade. "I hope you are right. There is a great deal of work to do, but I believe we can do it together. The trouble is that people are scared, and when they are scared, they will not help make things better."

"You're right, Doc. We need people to feel safe enough to help us clear out the bad guys. Then we can get to work putting everything back together. You say people are scared, but you didn't seem too scared today. Fingering those two guys in broad daylight was pretty ballsy."

Dr. Manza shook his head and spread his hands in a gesture of humility. "It was not so fearless. They were bad people. Good people must do something, or things will never change. I hope that my example will help them be less afraid."

"I think it will help, Doc. At least, I hope it will."

Our discussion wandered back and forth in this manner for a bit. Dr. Manza was a good conversationalist, and he had a good sense of the neighborhood and what needed to be done to get it put back together.

"The school must be opened, Dan. The children must go back to school and stay off the streets. Also, people must be able to go back to work, and we must restore electricity to their houses."

"I agree with all of that, Doc. Unengaged, uncomfortable people are bored and irritated, and bored and irritated people cause trouble. We also need to get the neighborhood cleaned up. The trash makes it too easy to hide explosives, and it creates the sense that nobody cares about the area. All of that is easier said than done, though. My infantry company is a little short on general contractors and electricians; do you know anyone who wants to rebuild the school?"

Dr. Manza took a drag on his cigarette, exhaled slowly, and looked me in the eye. "Yes, Dan, I do know someone who can help with those things. I have a friend who does that sort of work, and we can speak with him about it if you like. Is there money to do it?"

There was money to do it. In fact, there was a *lot* of money, but most of it wasn't mine to give him. I was sure my commander, Captain Hamilton, would be willing to support the effort, but I'd need to get him to talk to Dr. Manza himself to work out the details. Also, there would be numerous bureaucratic hoops associated with the government contracting process to jump through, quotes to gather, and paperwork to complete. All of that could be quite painful. I told Dr. Manza this, and he listened thoughtfully as he smoked. As an Iraqi who had prospered under Saddam's regime, he understood how the wheels of bureaucracy turned, and, in hindsight, I think he understood more about how to navigate the military contracting process than I did.

We continued talking for another ten minutes or so, until Dr. Manza started to get a far-away look in his eye, and I got the sense he wasn't paying much attention to the conversation anymore. I stopped talking, and after a few seconds, his point of focus changed as he noticed the silence.

"Sorry, Dan, I was just thinking about something else."

"What's up, Doc?"

Dr. Manza opened his mouth for a moment as though he was going to say something, but then he half-smiled, puffed on his cigarette again, and shook his head. "No, it is nothing, Dan."

I thought about pressing him, but instead I left Dr. Manza to his thoughts. The conversation didn't pick back up, so I looked at my watch and decided it was time to head back outside.

"Well, Doc. It's about time for me to leave. I'll talk to Captain Hamilton about our conversation and see if I can get him to come over and talk through the details with you."

"Thank you, Dan. I would appreciate that."

I leaned forward on my rifle butt and wrestled my way out of the couch's grasp to stand up, taking off my glove to shake Dr. Manza's hand.

"I'll see you soon, Doc. Thanks for the help today."

"Goodbye, Dan. Thank you as well."

Mr. Fasil showed me out the door and smiled as I walked back to Van Awesome's truck. I climbed in and told Van we were finished here.

"Anything hot happen while I was inside?"

"No, sir. Kaluzny and Gonzo are back, and they're RIPing with 2nd Platoon on Main Street now.

"Great. You ready to limp our wounded truck back to Falcon?"

"Yeah. We need to get it to the motor pool and pick up a new mine roller. I'm also going to order McDowell, Eubanks, and Tinney to get checked by the med section for any signs of concussion after that blast. I've heard about guys who were fine right after the blast but had trouble later."

"Yeah, that's a good idea, Van."

As we rolled back to Falcon, preceded by the tortured squeal of the damaged mine roller, Van's comments echoed through my mind. Had I made the right decision? Were these guys going to have permanent brain damage because I wasn't willing to wait for EOD?

I shunted these thoughts to the back of my mind and flipped open my notebook to scribble down notes on the meeting with Dr. Manza. I needed to debrief the Commander when I got back so he could meet with Doc, and I needed to make sure I had my thoughts in order.

Could Dr. Manza be the guy to help us straighten this place out? He certainly had the *cojones* for it. If he didn't get murdered, and if he really did have the necessary connections to help rebuild the neighborhood, it was possible he was *the* guy we'd been looking for to help us turn the neighborhood around.

Intermission

"Would my wife still love me? Would my son even remember me?"

Those were the recurring thoughts that ran through my brain as I headed home for R&R in September, 2007. R&R, or rest and recuperation, is slang for the free vacation time the Army provided to every service member who served a year-long tour in Iraq. The vacation wasn't counted against the normal thirty days off soldiers get every year, and it came with a free plane ticket to anywhere on the planet. Guys used this leave for a variety of purposes, from being present for the births of their children to hedonistic romps in Amsterdam. But for me, it was a chance to fly home and see Alycia and Gareth.

On the surface, R&R seems like it would be a relaxing experience–a little oasis of normalcy in a year long trek through the desert–however, the reality was somewhat different.

I flew home in the cleanest uniform I could find. It had no blood stains, relatively little discoloration, and an intact crotch. The trip home was pleasant enough, although the German customs officials confiscated my pocketknife when I transitioned through Frankfurt because the knife was illegal there.[1] Throughout the entire journey, all I could think about was seeing Alycia and Gareth. I hadn't devoted nearly as much time and energy to maintaining our relationship as I should have; there was just never the time or mental bandwidth to call or write, and now I was worried I might have done permanent damage to the relationship. Would Alycia and I still get along? Would it feel weird when we were together? Would it be hard to be in the moment, or would I keep thinking about Iraq? What about Gareth? I hadn't seen him since he was nine months old. He

---

1. German customs officials have since confiscated a half-dozen more knives from me. It seems like every time I fly through Germany, I have a pocketknife in my luggage somewhere that gets replaced with an incomprehensible note.

was a year and a half now and walking and talking. Would he be afraid of me? Would he remember me at all? I was simultaneously nervous and excited about meeting my family, and I worried about it the whole way home.

When I landed at the Kansas City airport, Alycia and Gareth were there waiting for me. It was a sunny and warm day, and as I walked out of the airport, I saw Alycia and Gareth next to our blue Toyota Camry. Alycia was smiling and beautiful, and Gareth was sitting in an umbrella stroller with a mop of curly blonde hair and a concerned look on his face. I closed the distance between us, wrapped her up in a hug, and kissed her. The kiss was long and wonderful, and after it was over, we stayed in each other's arms and looked at one another.

"Hi babe, it's good to see you."

"It's good to see you, too."

I eventually broke our embrace to squat down in front of Gareth. I'd bought a stuffed camel on the way home, so I handed it to him and smiled. He smiled back at me shyly and said, "Hi, Daddy." I took his hand and said, "Hi, Gareth. It's good to see you." I figured a hug would be too much at this point, so I turned back to Alycia.

"You ready to go home?"

"Yes. Yes, I am. It's *really* good to have you back."

"It sure is."

The drive back to Fort Riley took about two hours. Our conversation was the sort of easy catch-up conversation that two people who know each other well have after a long separation. My fears about whether we'd still get along were completely alleviated as we talked about family, friends, and the minutiae of our last few days. Those first few hours of conversation felt completely normal.

When we got home, I walked in and started to put my bag down where I always had. There was a running stroller there now, though, so I looked around and set it somewhere else. Alycia noticed and said, "Oh, I reorganized the entryway. Your bag goes over there now." I moved my bag and walked into the kitchen.

"Wait, would you mind taking your boots off before you go in?"

We'd always taken our shoes off before walking through the house, but I had completely forgotten. "Of course, my bad." I took off my boots, put them on the rack, and walked into the kitchen. I had that sticky post-travel feeling and wanted to get cleaned up.

"I'm going to hop in the shower real quick. That alright with you?"

"Of course, I'm going to put Gareth down for a nap."

As I walked through the living room, I was blown away by the number of toys I saw. The room was swimming in them. I wound my way through the room and made my way to the bathroom. I stripped down, turned on the water, and stepped in, inadvertently hitting my head on the ceiling; I'd forgotten how short the shower was. I grunted, soaked down my hair, and looked for my soap. It wasn't there. The shower setup looked different, and none of my stuff was there. I used Alycia's soap instead, finished my shower, and got out.

After I stepped out of the shower, I looked at myself in the mirror. I hadn't noticed the changes to my appearance over the past several months, but in the context of my old Kansas bathroom, I noticed the differences. Ten hours a day in the blazing Iraqi sun had tanned the lower half of my face as dark as it'd ever been. My neck was similarly dark, as were the strips of exposed skin between where my uniform sleeves ended and my gloves began. The rest of my body was pale white. Months of hard physical activity, dehydration, and weird nutrition had left my physique stringy. Even my eyes looked wrong.

I looked for my toothbrush, but it wasn't there either. Gareth's baby toothbrush was in its place. "I don't belong here. This isn't my home anymore." The thought entered my head unbidden. I dismissed it immediately-of course I belonged here. Things were just weird because I'd been gone for so long. I wrapped myself in a towel and walked from the bathroom to our bedroom. I opened my dresser and reflexively grabbed a brown T-shirt, but then I laughed at myself, put it down, and grabbed an old college T-shirt instead. It didn't fit right. Neither did the shorts. My mismatched tan looked especially weird with civilian clothes. I walked downstairs and found Alycia on the couch.

"Gareth's asleep–he won't be up for an hour or two. I was thinking we could go to Famous Dave's for dinner tonight if that sounds good to you."

"That sounds great. I haven't had good barbeque in forever."

We talked for a while, continuing to catch up on everything we had been doing for the last seven months. We weren't uncomfortable together, but our conversation had the stilted quality of two people who were friends once but didn't have common interests anymore. Alycia had been completely immersed in being the stay-at-home mom of a toddler. Her life, routine, and experiences revolved entirely around that. My life had been focused on something else. We'd each been living intensely focused but entirely separate lives, and now that I was back in her world for a while, it was hard to find common topics of conversation.

Gareth eventually woke up, and I played with him for a bit before dinner. So long as I played with toys and read books with him, we had all the common interests necessary to have a great relationship. It seemed like the Daddydoll and videos had done their job pretty well, and it didn't take long before Gareth was playing with me like I'd never left. We eventually piled into the car and drove to Famous Dave's. Alycia had been the one to drive home from the airport, but I decided it was my turn to drive this time. As I shifted the Camry into gear, I turned to her and said, "If I blow off any major traffic laws, let me know. It's been a while since I had to follow the rules of the road."

She smiled nervously. "Okay. Are you sure you're okay to drive?"

"Yeah, I'll be fine. It's like riding a bike, right?"

The drive was a strange experience. I failed to notice the first few traffic lights until the last minute. What few traffic lights were in *mahala* 838 didn't work. Even if they had, we wouldn't have stopped at the intersections. We owned the road. People stopped for us, not the other way around. While I was sitting at lights or stop signs, every piece of trash or casually walking pedestrian caught my eye. Were they threats? My knuckles whitened as I squeezed the steering wheel. I also didn't like having other cars around ours. It felt wrong. Who were these people, and why were they driving so close to me? Were they going to shoot at us? Why hadn't they pulled off to the side of the road as I pulled up to them? Was one of them going to explode as I passed?

None of that happened, of course, and we made it to Famous Dave's without incident. The hostess seated us at a booth, and we got Gareth a highchair and a sticky placemat to color on. The restaurant was crowded. There were people everywhere, talking noisily and walking past our table. The waitress brought ice water and our menus, and I looked at mine to figure out what to eat. I looked up at Alycia, who was smiling at me. She clearly wanted me to relax and have a good time, but I was having trouble tuning out the ambient noise and the people moving around me. I felt bad and tried to make conversation, but again it was stilted and a bit awkward. Everything was fine. Why was I having trouble feeling normal? Again, the thought came into my head that I didn't belong here anymore.

Over the next week, things smoothed out. I started to relax more easily, and the jetlag passed. We took Gareth to the zoo, the park, and to get ice cream in Manhattan. Alycia and I talked more easily, and things started to feel more normal. Even driving stopped being a stressful event. One night, after Gareth was in bed, we were sitting on the couch reading together. I commented on how nice things had gotten.

"You know, I was worried coming back home that I wouldn't be able to readjust very easily. I've had the sense that I didn't belong here a couple of times over the last few days, but that's fading too. I think things are going to even out just fine. Sorry if I've been a bit distant."

"No, it's okay. I understand. It's been a bit weird for me too. I'm glad things are feeling more normal. It's a shame you have to go back in less than a week."

And there it was. Just when I was starting to feel normal, I realized it was almost time to go back to Iraq. After that realization, it became impossible not to see everything we did in terms of how little time I had left at home. When my friend Ed came to visit, the whole trip became an all-consuming math problem. "When he arrives, we'll have four days of leave left, but by the time he's gone, we'll only have two days left together. That visit will use up 50% of our remaining time together." Everyday tasks became similarly quantified. "We're eating pancakes for breakfast together. We probably only have one more pancake breakfast together until I redeploy," or, even worse, "This might be the last pancake breakfast we ever eat together at all, because I might die in Iraq."

R&R was a battle to fight off the tendency to dwell on the amount of time we had left together. It became the elephant in every room. Worse, redeployment preparations started to consume our time together as well. I had to call and make reservations for the flight. I spent an hour on the phone talking to the agent while Alycia played with Gareth. I had to get my name and rank sewn onto a few new uniforms to replace the ones that had been ruined. That took another hour, which we could have spent doing something else. Alycia was patient about all of this, but redeployment preparation was a rival for time I needed to spend with her and Gareth, which made me feel guilty and made everything awkward.

By the time Alycia dropped me off at the airport, we both agreed that R&R was a bad idea. Just as we'd started to grow back together and enjoy our time as a family, we'd started preparing to separate again. Moreover, we both knew the upcoming separation would be at least as long as the first one. Because of the three-month extension, we were barely half-way through the deployment.

As I hugged Alycia and Gareth goodbye at the airport, I wondered if I'd ever see them again, and the walk to the plane was miserable.

On the plane, the apprehension of returning to combat overtook the misery of leaving home. What had I missed? How were my guys doing? Would it be the same when I got back? Combat soldiers are an insular bunch, and the distinction between "us" and "them" could be very subtle. I'd seen this frequently. A guy reassigned from one of the platoons to the headquarters section might still be on friendly terms with his old squad members, but after a few weeks of patrolling, a separation between them invariably developed. He wasn't part of the team anymore. He wasn't "us." I'd seen it repeatedly in Afghanistan, and I wondered if Iraq would be the same way. Was I "them," now? Would my relationship with the platoon be the same as it was before I left?

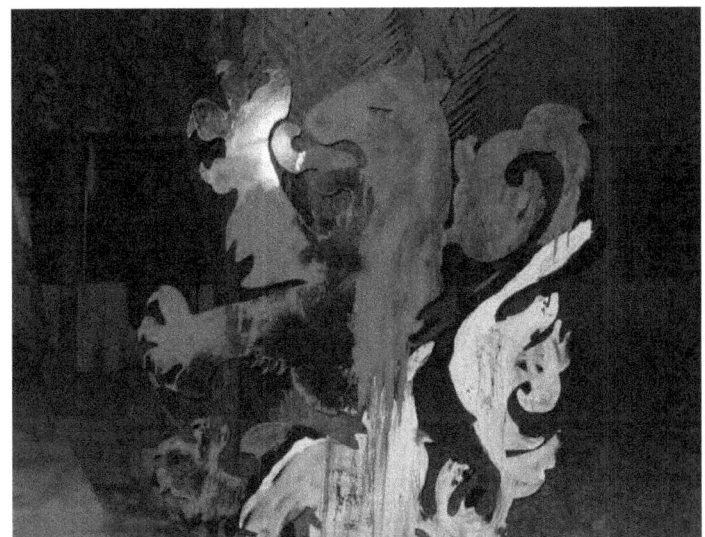

The aftermath of Operation Gato Negro

The courtyard of COP Amanche, the repurposed Catholic seminary

A propane distribution operation

Lieutenant Smith standing in an IED crater in Mechanics

The aftermath of an IED strike in Mechanics

The destruction of an enemy cache in eastern Mechanics

1st Platoon, Comanche Troop, under the Saber Arch Monument

Chaz (left) and Lieutenant Pace (right)

Lieutenant Pace (left) and SFC Kaluzny (right)

Main Street, Mahala 838

Comanche 13A, with SPARK roller

COP Banshee

McDowell (left), Orange Eid Bear (center), Van Awesome (right)

The Shakur Brothers' Car

Comanche soldier in remnant of an HBIED in Mahala 838

Raider Battalion memorial ceremony

Part IV

Combat Mayor

Grey Zone Ethics LLC

16

COP Banshee, Iraq, October, 2007

"I'll give you the *Belladia* and the Doura Refinery for Haji Kramer's house."

"Fuck you, no way. If you get Kramer's house, you'll be able to build a generator."

"So? You'll be good. Think of the cash you'll rake in with the *Belladia*."

"No. Go fuck yourself. Throw the dice."

"C'mon man, let's make a deal. What else can I throw in?"

At this point, the conversation was slowing things down, so the rest of the players chimed in.

"Hurry the fuck up."

"Throw the dice already."

"Cut the bullshit, man. I got guard in an hour."

"Alright, alright."

Schryver threw the dice and counted the squares as he moved his small, metal car.

"One, two, three, four, *fuck*! Sheik Ahmed's palace."

Doc Osborne cackled with delight. "That'll be 1,400 bucks, Sarge."

Schryver counted bills from his stack of cash until he realized he was going to come up short. He tossed his money and pile of properties to Doc and said, "I'm out, bitches. I need to go check on the guard towers, anyway."

The game of the day was *COPOPOLY*, a homemade version of *Monopoly* I'd designed in Microsoft PowerPoint, printed in secret on the Squadron plotter (to avoid the S3 Sergeant Major's wrath for using precious ink), and mounted to a box built out of an old ammo crate with Doc Osborne's help. The squares and cards had all been renamed for things from the *mahala*, and *COPOPOLY* became a favorite platoon pastime during our time at COP Banshee, our home in *Mahala 838*.

We lived at the COP three days out of every nine. When we were there, we manned the towers and the radio, creating a secure lily pad for the Squadron to use during its operations in Doura. The COP sat on the eastern edge of *Abu Tayara*. It was a small compound, no more than 50 yards across at its widest point, and it consisted of a two-story house, a few small outbuildings, and a T-wall fence with prefabricated concrete towers at the corners. As with almost every American base in Iraq, the entire COP was filled with fist-sized gravel to minimize mud when it rained and keep down the dust when it didn't. This gravel made walking anywhere a noisy, crunchy experience.

Guys crunched across the COP to climb the ladder to their guard towers for their four-hour guard shifts, then they crunched back across the COP to get Otis Spunkmeyer muffins or hot chow from the mobile kitchen trailer. They crunched over to the garage gym we'd built in one of the outbuildings, then they crunched back to their rooms in the main building to watch movies or sleep. Life was crunchy but simple at COP Banshee, and most of the platoon enjoyed their time there. The simple routine of eat-sleep-guard-workout-fuckoff-eat-sleep was a welcome counterpoint to the always-on, hectic grind of patrolling, and it gave guys the chance to internally process their combat experiences. There were as many decompression techniques as we had guys in the platoon, and it was surprising how many ways people found to occupy themselves in an area as small and austere as the COP.

For instance, Sproul was a born worker, so his version of relaxation was to buy a few sacks of concrete from one of the shops in *Abu Tayara* and pour a new paddock outside the entryway to the sleeping area to keep it a bit cleaner inside. Some days, as I made my rounds around the COP, I'd find him standing there for six or seven hours, cheerfully pouring, stirring, and smoothing out concrete. In Sproul's world, concrete was a known element. It was something you could control because concrete made sense. If you built a solid frame and poured in the right amounts of water and concrete, you got a concrete paddock. If you did a good job and were competent, you could produce a valuable object that served a discernible purpose. Our combat experience was nothing like that. Our inputs resulted in haphazard outputs, and it was often unclear what purpose any of our specific actions served. For a guy like Sproul, who valued order and accomplishment, pouring concrete helped him clear his head and anchor himself in something he understood and enjoyed.

La Claire and Walters decompressed differently. For them, the best way to cope with the deployment was to look past it to the awesome things they'd do when they got home with their pockets stuffed with combat money. COP time let them carefully plan the trim packages on the new trucks they'd buy, chat up the lovely ladies back home on *Myspace* or *Hotornot*, and work on building beach muscles. They spent their COP days crushing creatine, protein, and Jack3d, and knocking out marathon gym sessions, before retreating to the plywood internet café with a plate full of chicken. For guys who didn't have much meaningful interaction with the Iraqis and instead spent almost their entire waking lives behind a gun, alternating between extreme boredom and making split-second life or death decisions, looking forward to an amazing future became a way to make the present seem worthwhile.

Blanco was a photographer. He was a temperamental guy, who frequently wore on his sleeve the wild emotional swings combat often provoked. But when he was snapping photos or sifting through and editing them later on his computer, he seemed to be able to detach from the difficulties of the deployment and relax.[1] His pictures ranged from mundane shots of flowers, sunsets, and smiling children to sensational wartime photos, like his pictures of the enormous Doura refinery fire or IED wreckage, though I never had the sense he prioritized one type over the other. When he saw a person, object, or situation that inspired him, he captured it with his camera, and the process of doing so seemed to help him work through the emotions associated with his experiences.

Other guys had other ways to decompress. The platoon had video gamers, movie buffs, and porn connoisseurs, and it had callers-home and investment researchers. We even had a few readers. Most pastimes were individual in nature, and frequently I could walk around the entire COP without hearing anyone speaking at all. When guys spent almost every waking hour in sector talking to each other, talking to (or shouting at) Iraqis, or talking on the radio, the COP provided a rare opportunity to be alone. Guard shifts were pulled alone. Meals were mostly taken alone. Even our rooms, which were all shared with at least one other person, were modified using ponchos or locally procured blankets to construct little *ad hoc* fortresses of solitude.

The exception to all of this solitude was board gaming. In the headquarters area of the COP was a large table near the radio rack that enabled us to stay in touch with the headquarters at Falcon, and for a few hours a day this table became the center of social activity on Banshee. We didn't have a large collection of games; occasionally, a new game would arrive or an old one would be destroyed in the furious aftermath of

---

1. Indeed, many of the photos in this book were taken by Miles Blanco.

a contentious session, or thanks to spilled Ripit, but among our favorites were *Risk*, *COPOPLY*, and *The War on Terror–The Board Game*. In contrast to the COP's usual serenity, during gaming sessions, emotions ran high, and the walls frequently echoed with violent shouting when tragedy struck, or fortune smiled.

Today's game was no different. With Schryver out of the game, only Doc and McDowell remained, and as Doc held the Dark Blues, the Greens, and the Reds against McDowell's paltry collection of low-rent properties, Doc's position was definitely the stronger one. McDowell's only strength was that he was wearing the lucky balaclava of evil–a prop that came with *The War on Terror* and was believed to grant its wearer good fortune. Unfortunately, the ski mask's magic didn't work, and within a half hour, the game was over. McDowell handled his defeat gracefully, grinning, and congratulating Doc on the win.

"Ah well, I need to get ready for guard, anyway. You probably cheated, you redneck bastard."

"Of course, Sarge. How else would I have managed to beat such a talented group of opponents?"

I interrupted the banter.

"Alright, let's get this crap cleaned up. The boss is coming out in a half hour, and we need the table for a meeting."

Our new commander, Captain Hamilton, had been coming into sector frequently over the last month. His initial meetings with Dr. Hamza had gone well, and the initial project Hamilton had assigned to Dr. Hamza to see if he was a capable contractor–rebuilding a neighborhood soccer field-had been completed on time and to a high standard. After the field was finished, Dr. Hamza had even hosted an inaugural game between the Iraqi police and the neighborhood team, and the resulting celebration had been well received by the local citizens. Our role had been to secure the event and to provide a few distinguished guests. We'd even had a few guys take to the field in a limited capacity, although theater policy said they had to play in armor, which put them at a bit of a disadvantage and made them look a bit ridiculous.

The aim of today's meeting at the COP was to discuss the next steps for improving the neighborhood. At the appointed hour, Hamilton arrived with Lieutenant Keller, our company contracting officer. Lieutenant Smith, who was patrolling *Abu Tayara* with his platoon, also came.

When Hamilton arrived, he hung his armor and rifle on the green metal weapon rack in the corner of the headquarters area and walked across the room to stand next to a large dry erase board near the central table. Hamilton was an infantry officer who had been with the Squadron for only a few months, but my initial impression of him was positive. For starters, he actually patrolled with us and visited the COP regularly. The guy he'd replaced had done little more than watch *Deadwood* and *Rome* and attend to the Squadron battle rhythm events back on Falcon. Meetings were indubitably important, but to the guys who spent their deployment in sector getting blown up, a commander who seemed unwilling to patrol was hard to respect. Like Cook, Hamilton took the opposite approach and delegated most of the FOB Falcon work to his XO and First Sergeant. I'm not sure what this did to the Company's reputation with the Squadron staff, but it certainly endeared Hamilton to his guys.

Once we were all seated around the table, Hamilton picked up a marker and started scribbling on the board.

"Alright, gentlemen, here's where we are. Dr. Manza looks like he'll be a reliable lead for our efforts to rebuild the neighborhood, so we're going to invest heavily in him to make that happen. We're going to have three main lines of effort. First, we're going to hire crews to clean up the streets, repair all the damaged curbs, and paint everything. If we can reduce the number of places the insurgents can hide IEDs, we'll reduce the number of attacks, and if we can give the people something nice to look at, we think they'll be more interested in taking care of it."

Rich and I nodded. We'd thought similar things. Hamilton continued.

"Second, we've been authorized to start a microgrant program to stimulate small business development. At my level, I can accept a proposal from a local with a business plan and give him up to $10,000 to make his plan happen. We're going to run a few events in sector over the next few weeks to solicit proposals. While that's going on, we're going to have Dr. Manza rebuild the *Al-Najun* school and install playgrounds across the neighborhood. All of this should keep people busy during the day, allow them to start earning money, and keep them from having to conduct illegal activities to feed their families."

Rich and I exchanged a skeptical look and started to comment. Hamilton caught it and interrupted us.

"Yeah, I know what you're going to say. How will we scrub the proposals to see which ones are legitimate? How are we going to avoid wasting money on bullshit businesses, right?"

We nodded. That summed up our concerns.

"The truth is, I'm okay with a few bullshit businesses. If we see that somebody is taking the cash and doing nothing with it, we'll cut him off, but what we're really trying to accomplish here is to get people to work and to get them invested in the neighborhood. So, if we need to waste some money to make that happen, I'm good with it."

He knitted his brows and continued.

"If you really want to get down to it, as a percentage of what this war is costing, how much money will we be wasting, anyway? What does it cost to keep our unit deployed over here? How much does it cost to rebuild every truck that gets destroyed by an IED? Aside from the monetary costs, how much wasted money is worth a dead soldier? If a few Iraqis scam us out of a bit of cash, and that incentivizes those Iraqis to support us instead of the insurgency and saves U.S. lives, I'm all for it."

What Hamilton said made sense. What was somebody else's money to a bunch of twenty-something soldiers? It didn't come out of our pockets. It hardly felt like real money at all. What *did* feel real, though, were booby-trapped exploding houses and flaming trucks full of dead people. Those felt real. So did seeing dead Iraqi kids in the streets and craters full of body parts. I'd trade fake money for a safe neighborhood any day.

"That makes sense to me, sir. What's the third thing?"

After his last comment, Hamilton had lapsed into silence and was staring at the board in thought. My question pulled him back out of his head.

"The third thing, Dan, is to get the utilities turned back on. This country is too goddamn hot, and people who can't comfortably sit in their homes and watch TV during the heat of the day or cook themselves dinner are going to be pissed off. We can't repair the entire power grid, so we're going to build two generators in the neighborhood and pay people to run them. That should provide enough power to keep everyone comfortable."

He turned and pointed at the large map of the *mahala* on the headquarters wall. "I'm thinking we should put them here and here."

The locations he pointed to were the sites of the old generators that had burned down a few months ago.

"That makes sense to me, sir. What do you need from us, other than what we're already doing?"

Hamilton's response was cut short by the room suddenly becoming entirely dark and quiet. The Iraqis weren't the only ones with electrical problems. The COP's notoriously finicky generator had just died again. I fumbled through the blacked-out room for the door to find Doc.

"Generator's out, sir. One sec, I'll get it back up."

I left the office and walked down the narrow dirt hall to Doc's room. There was no door, but there was a poncho hung across the doorway. I pushed the camouflage fabric aside and looked inside. The room was completely dark, and Doc was lying on his cot in shorts and an Army-issued brown T-shirt with a portable DVD player on his chest and headphones on his head. The blue glow of the player's small screen illuminated his face. As I walked in, he took off the headphones and looked up at me.

"Wuzzup, sir?"

"Generator's down, come help me get it up."

Doc nodded and sat up. He stuffed his feet into his boots without lacing them and followed me down the hall and out of the building. We pushed open the exterior door and were momentarily blinded by the strong Iraqi sun. Sproul was standing in front of the door in flip flops and a t-shirt putting the finishing touches on a concrete pad he was working on.

"Don't step on the concrete. It's still drying." He cautioned us.

We nodded and stepped around the pad, crunching our way across the gravel to the side of the building where the generator sat, huge and silent in the afternoon sun. It was a yellow, 100-kilowatt diesel model which was sitting on a couple of old wooden pallets, and its electrical output was all that separated us from a hot, dark life of stifling misery. Getting it up and running again was of critical importance. Unfortunately, sometimes that wasn't easy.

One problem was that our generator was never intended to serve as a primary power source for an entire building in a 120-degree desert. Running it constantly, day after day, with no down time or maintenance and nothing but dusty air running through its filters took its toll. The other problem was that none of us were generator mechanics. The team that had installed the generator had left the manual, but most of us looked at the machine like the apes looked at the obelisk in *2001: A Space Odyssey*.

That's where Doc came in handy. In addition to being a great medic, Doc was one of those people who can fix just about anything. His ability to rig up something out of nothing made *MacGuyver* look inept. As we walked up to the generator, he looked at it for a few seconds, opened a few panels, and pushed some buttons. He pulled a Gerber multitool out of his pocket and pried off a piece of the generator.

"Give me a few minutes, sir. I'll get it working."

I left him to his project and walked back to the headquarters area. Hamilton and Smith had propped the door to the headquarters room open to let some light into the room and were sitting around the table talking. Hamilton acknowledged me as I sat down and rejoined the conversation.

"Rich and I were just discussing the state of the *mahala* and the way ahead. Assuming Dr. Manza remains reliable, we're going to leverage him as our lead on the neighborhood reconstruction effort. That effort will fail unless we keep the number of attacks down, though."

Rich spoke up. "We've got a pretty good network diagram of the bad guys in the neighborhood at this point. We're mostly just waiting for sufficient evidence to arrest them."

"Don't worry about getting too much evidence. When we get information on terrorists or criminals in the neighborhood, just take them off the street. Some of them will be prosecuted and put in jail. Some of the others will probably be released after a few weeks because there's no credible evidence. In either case, they'll be off the streets for a while, which means they won't be impeding our efforts to rebuild this place. Even if they get released, they'll be unlikely to want to cause more trouble and risk getting arrested again. Mow the lawn to keep the weeds down while we give the grass a chance to thrive."

I chimed in. "We've been locking up abandoned houses to keep them from turning into HBIEDs. We can expand that effort and lock up all the abandoned houses in the neighborhood, then only let people move into them if they can prove ownership. That'll keep the number of staging areas available for bad guys to a minimum."

Smith nodded. "Let's take that one step further. I've had several new families approach me on patrol, asking if they can move into the *mahala*. Most of them have relatives here. Let's let them move in. We can interview them, screen their packets, and put all their information into the neighborhood database."

I laughed, pulled out the huge ring of keys I carried to all the locks we'd put on abandoned houses over the last few months, and dropped it on the table. "We'll be combat real estate agents. Do we get a commission?"

Hamilton laughed too. "Maybe so. Six percent is the going rate, right?

My face hardened as a thought crossed my mind. "I guess maybe our commission will be staying alive for the rest of the deployment."

Smith and Hamilton nodded. Sometimes it was easy to get so focused on the discussions about urban renewal and abandoned houses that I forgot what the stakes in this game really consisted of. At that moment, the lights popped back on, and the air conditioner whirred to life. Hamilton pushed back from the table and stood up.

"On that uplifting note, gentlemen, I think we're done here. Rich, I'm going to head back into sector with you and have a meeting with Dr. Hamza this evening."

Rich nodded, and they both started putting on their gear. Hamilton turned to me, sweeping his arm around the room.

"Dan, once again, all of this is yours."

As he and Smith opened the door to leave, I heard the rhythmic scrape of Sproul's shovel as he continued to smooth out the fresh concrete of the pad he was working on.

17

Mahala 838, Iraq, November, 2007

"Hey, One-two. How many is that for your squad?"

"Seven, Sir"

"Roger. One-three, how are you looking?"

"We've got six so far. One-three-alpha is working on the seventh now, but the tires are all flat, so it's taking a bit longer than the last few."

"Roger. Thanks."

I unplugged my headset from the radio and got out of Blanco's truck to see how he and Cisneros were doing with the car they were working on. It was a pleasant November night, and *Abu Tayara* was quiet. Our mission this evening was to clear abandoned cars out of the neighborhood.

The car Blanco and Cisneros were dealing with was a white Daewoo Prince that had seen better days. It had one door that was painted primer gray, a broken passenger side mirror, and a muffler which dragged the ground. The rear bumper had the Daewoo logo that all Princes shared, but its owner had decorated the front grill with the rearing horse of a Ferrari and emblazoned the driver's side fender with a cobra taken from a Ford Mustang. We'd marked the car a few days ago as one we'd suspected of being abandoned, but as the Ferrari-Mustang-Prince was still here, it was clearly unloved and thus destined to be towed. Getting rid of all the abandoned cars in the neighborhood was part of our effort to clean the place up. The curb repairs were mostly complete, the effort to clean up all the rubble from the various IEDs and HBIEDs was underway, and the guys we'd hired to clean up all the trash in the streets were making good progress.[1] The cars, though, remained. They cluttered the side streets, creating traffic problems and ruining the *feng shui* of the *mahala*. This would have been irritating but acceptable, but as they also provided an excellent place for interested parties to conceal IEDs, the cars had to go.

As I walked up, Cisneros was jimmying open the door with a hooligan tool so he could pop the emergency brake and shift the transmission into neutral. Blanco was maneuvering his truck behind the car to get ready to push it when Cis was ready. The streetlights were working tonight, so none of the crew were using night vision.

"Need any help, Cis?"

"Nah, sir. I almost got it."

I stepped back and let Cis do his thing. Cisneros was one of our newly minted sergeants. He was a quiet, cheerful guy with a penchant for opening all kinds of doors and windows. Over the months, he'd jimmied windows, jumped over gates to unlock them, and shot the locking mechanisms out of a plethora of front doors with the Mossberg twelve-gauge shotgun he kept in his truck. I don't think he had any formal training in the subtle art of getting into things; he just had a knack for it.

It took Cis only a few moments to finish, and the car rocked forward a bit as he released the brakes and slipped the transmission into neutral. Cis stepped back with a satisfied look on his face and yelled to Blanco, "Alright man, it's ready."

Blanco was standing next to his driver's door. He gave Cis a thumbs up and leaned into the open window of the truck to tell his driver to start moving. The diesel engine grunted as the truck lurched into gear and started to roll forward, headlights flipping on to bathe the Prince in white light. Blanco walked alongside the truck, and Cis used his hands to guide them forward until the front grill of the Humvee made contact with the rear bumper of the Prince. Cis gave Blanco a thumbs up, then climbed into the Prince to steer it as the Humvee accelerated to push the Prince down the street. The Prince's poorly maintained suspension groaned as it bounced up and down under Cis's armed and armored bulk.

---

1. I was initially surprised when a trio of middle-aged men applied for the trash removal contract. I'd seen forty-something men running shops, drinking chai, or driving cabs, but I'd never seen one doing manual labor. For a few hours, I thought maybe our work ethic had rubbed off on the locals a bit. The next day though, when I saw one of our new employees sitting in a lawn chair and smoking a cigarette while he managed a crew of a half-dozen small children picking up the trash for him, I understood my mistake.

The car rolled smoothly along the street, bumping along on its mostly flat tires. Occasionally, the car would pull ahead, creating a small separation between the two vehicles, and causing a solid thud when the truck regained contact. As his helmeted head rocked back and forth with the jostling, Cis grinned and loudly whistled the tune to *Low Rider*. He guided the Prince to the end of the block, turned south, and led the convoy bumping and thumping to the dirt lot near the edge of the *mahala* we'd designated as the neighborhood junkyard. Cis parked the car, not the simplest task with a dead engine, then got out and climbed back into Blanco's vehicle to continue the patrol.

This sort of mundane work was critical to our success in the *mahala*. It wasn't sexy, and it didn't brief well when we reported to higher, but car removal was critical to keeping my boys alive. As I looked at the dozen vehicles in our improvised junkyard–adjacent to the southern police checkpoint so the Iraqis could keep an eye on the cars–I felt a sense of satisfaction. Each of these jalopies could easily be converted into a VBIED, and a single VBIED could kill half of my platoon, destroy one of the Iraqi police checkpoints, or wreck the gate of FOB Falcon. Trash removal wasn't something the Army put in recruiting commercials, but it was important work in our counterinsurgency fight.

"Comanche One-six, this is One-three. One-three-alpha has a situation up here across from the school, and we need you to take a look."

"This is One-six. I'm on my way."

I could see Van Awesome's trucks a few hundred yards north of where I was standing, so I told Blanco's squad to continue their work without me while I walked north to link up with Van. The streetlights were partially working, but it was dark enough that I kept my night vision goggles down. The walk was short and free of the sewage, IED craters, and feral dogs that had been common obstacles during our earlier night patrols. There were still parts of *Abu Tayara* that were in bad shape, but our initial efforts had done a pretty good job of getting the main street cleaned up.

When I arrived at 3rd Squad's location, I found their two trucks pulled off to the side of the main street with their turrets facing out and their crews pulling security. The ten-foot walls of the *Al-Najun* school were just east of the trucks, and between them sat a brown '80s Oldsmobile Cutlass. There was nobody inside the car, but Van Awesome was standing next to it. As I walked up, he turned toward me, flipped up his night vision goggles, and updated me on the situation.

"Sir, McDowell has two guys balled up in that house over there." He gestured to an unobtrusive, gated home across the street from where we were standing. "While we were looking for cars to tow, two guys came flying up the road in this Cutlass. They had the windows down and shouted something at us that sounded assholish, so I had the other truck cut them off so we could talk to them. When the car stopped though, the two guys got out, ran off, and jumped the gate to that house. McDowell, Sproul, and LaClaire chased them into the yard, but the two guys charged McDowell as he came through the gate, so my guys beat the hell out of them. That calmed them down, and now they are talking to Ozzie."

"What are your thoughts, Van? Were they out here putting in IEDs tonight, or are they just a couple of local dumb asses?"

"I imagine if they were looking to kill us, they would have been more subtle, but who knows?"

"I guess we'll find out soon."

Van and I walked through the gate where LaClaire was pulling security, and the sight that greeted us as we entered the yard took me back to my college days. Sproul had the flashlight of his weapon trained on two young Iraqi men in tracksuits. They were sitting with their backs against the wall with their hands flex-cuffed behind them. Their eyes were bloodshot, their lips were bloody and starting to swell, and they were squinting hard against the harsh, white light. McDowell and Ozzie were standing outside the glare of the flashlight, asking them questions. It looked like a scene from *Cops*.

McDowell's voice was harsh as he shouted his questions. "Who are these mother-fuckers, and where do they live?"

There was a pause as Ozzie translated, then another as one of the men slurred his response.

"Their names are Ahmed and Mohammed Shakur. They say they are brothers and that they live on 16th street with their parents, but that they've been out of the area for the last year to avoid the trouble."

"Give me their IDs." On Ozzie's translated command, the two produced laminated ID cards, which were the Iraqi equivalent of driver's licenses.

"Their information checks out. They're registered as residents of this *mahala.*"

"Why did they come back to the neighborhood?"

There was another pause while they spoke in Arabic. "They heard the neighborhood was good again and that they could get jobs here."

"Okay. Why were they being such assholes then?"

Ozzie didn't even pause to translate. He turned to McDowell, "They're drunk, man. Look at them, they're a wreck, and I can smell them from here. I don't think they're terrorists. They're just idiots."

McDowell blinked for a minute. His blood was up from the fistfight, but as he listened to Ozzie, he started to calm down, and in a few seconds, he started laughing at the ridiculousness of the situation. He turned back to the two Iraqis and said, "You turds need to sober up. Give us your keys and walk home."

Ozzie translated. The two guys looked confused for a second and then both nodded. McDowell and Sproul jerked them to their feet and cut off their flex cuffs. The men smiled nervously as McDowell started lecturing them, his tone level and amused. "You two idiots could have been *killed* tonight. There are people burying IEDs and executing people every night in this *mahala*, and we almost shot your dumb asses because you were being stupid. I like getting drunk as much as the next guy, but you need to appreciate the gravity of what's going on in this neighborhood. I'm going to let you go with a warning, tonight. I'll park your car at the police checkpoint and leave the keys in the ignition. If I see you messing with it before tomorrow morning, I'm turning you over to the Iraqi police, and they'll take you to jail. Now, go home and sleep it off."

I'm not sure how well McDowell's statement translated into Arabic, but once the Shakur brothers heard it, they stood up, brushed themselves off, and left the yard. Heads hung in shame, they walked north up Main Street, presumably to their homes, and we didn't see them again for several days.[2]

Once they were gone, McDowell turned to me and Van, swinging the keys to the Shakur brothers' Oldsmobile around his finger.

"Who's up for a drive in a Cutlass?"

I smiled. "She's not going to park herself. Let's get her up to the police station."

The three of us walked out to the car. As we walked, I talked to McDowell about the scene I'd just witnessed.

"Man, it's good you didn't shoot the Shakur brothers tonight. I'm surprised you didn't gun them down when they jumped you. Kudos on the self-restraint."

---

2. Once we did see them again though, they started turning up everywhere, all the time. The Shakur brothers became some of our best and most entertaining friends in the neighborhood, and we had many hilarious conversations with them over the next several months.

McDowell looked down at his hands. He was wearing shooting gloves that had reinforced Kevlar knuckles for punching people. They were scratched up and bloody. "I don't know why, but when the first dude jumped me, my immediate instinct was to crack him in the face. I probably should have gone for my rifle, but I just did what came naturally."

"It's good that you did, man. I don't think our new friends in the neighborhood would have been too happy if we'd killed a couple of their kids over a misunderstanding."

Van chimed in. "On the other hand, if they'd had weapons and wanted to kill you, you'd probably be dead."

McDowell chewed on that response for a minute. "You're right, Van. How many things over here are like that? You have a split second to make a decision, and you have no real way of knowing whether you made the one that will keep you alive, the one that will get you killed, or the one that will screw up the whole mission for everyone. I feel like we've been really lucky."

I nodded. "Yep. I guess it's better to be lucky than good, right?"

"Yeah, maybe so."

We arrived at the Cutlass. It looked like an '80s model to me, and it was painted a well-worn, rusty version of that shade of chocolate brown that apparently seemed like a good idea to car manufacturers back then. Unlike most of the cars we saw, the Cutlass hadn't been stickered up, bedazzled, or rebadged. It was just a straight up '80s beater which had somehow found its way to Iraq.

McDowell opened the driver's side door and sat down on the tattered velour seat. The car immediately sagged to the left under his weight, and his flipped-up night vision goggles scraped the roof. He shut the door, cranked the car, and rested his elbow on the open window.

"Whaddya think, sir? How do I look?"

"Pretty bad ass, although you need a Coors Light and a trucker hat to complete the image.

"Why don't you guys climb in here? Let's take her for a spin."

"Now that's a damn good idea. I don't remember the last time I rode in one of these. You coming, Van?"

"Why not?" He turned to his truck where Schryver was standing. "You're coming too. Get in."

We opened the passenger door and piled into the cutlass. It was a two-door, so it was a bit snug, but after a deal of squeezing, untangling, and swearing, Van was in the passenger seat, Schryver was in the back, and I ended up hanging out of the sunroof. McDowell shifted the car into gear, and we rolled north. Under our combined half ton of weight, the car's fenders were dangerously near to scraping the tires.

"Give it some gas, man! Let's see what this baby's got."

McDowell hit the accelerator, and the engine roared to life, propelling us up the road at forty miles an hour. McDowell had flipped his night vision back down and turned the headlights off, so we were driving blacked out. After a few blocks, he locked up the brakes and jerked the wheel to see how the Cutlass handled, which as it turns out, wasn't very well.

"That's probably far enough. I don't want the police to light us up with a machine gun if we drive up to their checkpoint in this thing."

"Yeah, that's probably a good idea."

We stopped the car several blocks south of the northern checkpoint, and I called the trucks on the radio to have them meet us. Van, Schryver, and I got out of the Cutlass. McDowell flipped on the headlights to avoid surprising the police, and 3rd Squad escorted the car the rest of the way to the checkpoint. They parked the car near the concrete serpentine at the checkpoint's entrance and left the keys on the front seat. Ozzie explained the situation to the Iraqi policeman who was on duty, and he gave us a thumbs up. McDowell got out of the car and walked back to his truck.

"Do you think we've seen the last of the Shakur brothers, sir?"

"Not likely. With all the families we've been letting move back into the *mahala*, I'm sure there are going to be a bunch of similar stories. I thought Muslims didn't drink, but the young dudes here seem to have no problem with it."

The Islamic radicals in Afghanistan I'd encountered had been relatively puritanical, but the bad guys in our neighborhood mostly seemed like criminals and opportunists with a penchant for bomb-making.

"With all these young guys moving back to the neighborhood for work, we're going to need to make sure they get jobs. Nothing is more likely to cause problems than a bunch of bored young men."

"That's a fact. Idle hands and all."

Idle hands, indeed. We'd interviewed dozens of families over the past few weeks, and after screening them for criminal records, we gave them the keys to an abandoned house in the neighborhood. We didn't have any documentation on who actually owned the houses, but then again, we didn't really care. Houses that were full of families were houses that weren't filled with explosives or terrorists, and we weren't sure how well property rights had held up through the invasion and regime change, anyway. From our point of view, the Iraqis could figure all that out later.

Most of these families had kids, many of whom were teenage males. The families wanted what families everywhere want for their sons—a path to success and independence. Unfortunately, this wasn't always easy to find in wartime Iraq. While our microgrant program was gaining traction, and businesses were opening, many of the new businesses were shops or small, café style restaurants that didn't need many employees. This situation left hundreds of young men out of work and bored. Many of them, like the Shakur brothers, turned to booze and shenanigans, but insurgent groups were always recruiting, and plenty of young men turned to violence and crime to make money. In the last month, in our effort to calm the neighborhood, we'd arrested hundreds of both sorts, some for serious crimes, but many for nothing more than looking suspicious or giving us the stink eye. We always followed up our arrests with a visit to the family to explain why we'd picked up their sons, and most of the arrestees were released after just a few days in U.S. custody. Once home, the delinquents were almost always straightened out by their families, and while we had a few repeat offenders, we didn't have too many. Still, the problem of what to do with all the bored young men in our neighborhood lingered.

McDowell, Van, and I were still chewing on this problem when we heard the rumble of diesel trucks coming through the serpentine behind us. Since all our trucks were in sector, this surprised me. Our relief wasn't scheduled to arrive for another few hours, so who were these guys? As though responding to my inner monologue, Kaluzny's voice came across the radio.

"One-six, this is One-seven, Raider Six is coming into sector and wants to link up with you. Raider Seven is with him and wants to talk to me, too."

"Roger, we're at the northern checkpoint now. Let's do the link up here. I wonder what's up."

"No idea. I'll come your way. Everybody, check your uniforms. Especially you, One-three."

Raider Six was Crider's call sign. He, Sergeant Major Jones, and the squadron operations officer, Major Callahan, periodically rotated through the Squadron's sector to talk to the platoons, meet key Iraqis, or escort the occasional distinguished visitor who visited our area. Despite the occasional friction between the sergeant major and my NCOs on uniform discipline, these visits were generally welcome, as our patrol schedule was so busy that we rarely had any other opportunity to give the squadron leadership our version of the ground truth directly.[3]

As Van and McDowell's guys fixed their uniforms, Raider Six's convoy rumbled through the checkpoint and stopped in front of us. The Sergeant Major's truck continued south to find Kaluzny, while Crider and Callahan got out of their trucks and walked up to talk to us while their security detail fanned out to form a perimeter around the area.

Crider's demeanor was as calm as ever. Even in the heart of Baghdad, his manner of speaking was the same as it was during a normal Wednesday staff meeting.

"Hi Dan, how are things today?"

"They're good, sir. It's been a busy day, but things have slowed down a bit since it got dark. We found a couple of Coke bottle IEDs earlier, but they were poorly built and hastily installed, so we cleared them internally.[4] The locals tipped us off on the IEDs' locations and asked us to clear them without blowing out all their windows."

He nodded. "Maybe the people are seeing us as the solution to a problem caused by insurgents, rather than seeing the insurgents as a solution to problems caused by us. That's good progress."

"Yeah, that sums it up pretty well. People have started to trust us enough to flag us down and point out problems they want us to resolve. Not too long ago, they would have been too scared to do that because they would have been worried about getting

---

3. In case I haven't made it clear, Iraq was *hot*. Consequently, guys hated wearing gloves, helmets, and safety glasses, and they often cuffed their sleeves and pant legs to cool down a bit. This was anathema to sergeants major across the theater. In part, this was because not wearing protective equipment led to people getting injured or killed, but it was also because senior leaders saw flagging uniform standards as an indicator of overall lax discipline. It was sort of like Rudy Giuliani's broken window policy.

4. We didn't discuss the specifics of how we'd cleared the IEDs. Because of the EOD backlog, we'd developed a controversial technique called the 'five-finger recon' to investigate potential IEDs. In short, one of us would reach into the suspected IED location, use a hand to roll it out into the open, and if necessary, pull the wires out of the top of it. Then, we'd take a few pictures and throw the disabled device in a sack to give to the Iraqi police. In hindsight, this was a stupid practice, but at the time, it seemed better than waiting all day for EOD.

executed by terrorists."

"At the squadron level, we're seeing that play out in the metrics we collect. Enemy initiated attacks are down substantially, and successful arrests are up. We're taking enough bad guys off the field to really swing things in our favor, and that's translating into both fewer U.S. casualties and the emergence of locals who are willing to work to improve their neighborhoods."

As a platoon leader, it was sometimes hard to see the forest for the trees, so hearing that there was discernible progress at Crider's level was very welcome.

"That's good to hear. It'd definitely be nice to not get blown up anymore. What's happening with all the guys we arrest, though? I've seen a few of them around town after they were released. Those are usually the guys we didn't have too much evidence on, though–the ones who were only taken in because angry neighbors complained about them. What about the *really* bad guys? Is the Iraqi justice system dealing with them?"

Crider's brows knitted as he responded. "Unfortunately, I don't have a lot of visibility on what happens to them after they leave the DHA. I suspect that once they leave U.S. custody, the Sunni criminals get treated harshly because the Iraqi justice system is mostly managed by Shi'a personnel. Conversely, I imagine the Shi'a criminals get fairly lenient treatment. To be frank, they're probably getting released with no harsher punishment than instructions not to come back to Doura. The Iraqi government understands exactly how to exploit the gaps and seams in the U.S. military's intelligence system to ensure we can't prove what they are doing."

I nodded. That didn't surprise me. Even at our level, the difference between how Shi'a and Sunni prisoners were treated by the Iraqi police when we turned them in was stark. Why would that disparity be any different at the national level?

Crider continued. "That's all above our level, though, and there's nothing we can do about it. As a squadron, we'll continue to focus on controlling what we can, and that will have to be enough."

Crider's eyes drifted over to the abandoned Cutlass. "What's the story with the Cutlass? You don't see many of those around here."

I recounted the story of the Shakur brothers. Crider listened to the whole story, including McDowell's enthusiastic interjections, with his characteristic patience. When we had finished, Crider replied.

"You handled that situation well. De-escalation was definitely the right answer. Have you seen a trend of friction between your platoon and the young men in the neighborhood?"

"They're usually the people we get reporting on, so they're usually the ones we arrest. Even the young men who aren't getting arrested are a problem, though. They're bored, and boredom leads them into trouble."

Crider nodded. "That's a problem everywhere. Especially in the Sunni communities. The young Sunnis can't get jobs in the army or the police, which are the organizations a lot of these kids would have worked for under the old regime, so they are easy targets for radical groups who offer them money to fight the foreign invader. I think there might be a solution though. Over the last several months, the senior leaders in the Green Zone have worked out a deal with the tribal leaders in Ramadi to work together with U.S. forces against the terrorists hiding in the Sunni communities. Thanks to this deal, the U.S. battalions in Ramadi have been authorized to recruit young Sunni men, vet them, train them, and employ them to secure Sunni neighborhoods. They're calling these units the Sons of Iraq. So far, these units have been much more successful than the Iraqi police at getting information and securing their *mahalas*."

Crider gestured toward the Cutlass with his head. "What do you think about implementing that strategy here?"

I thought for a minute. "My initial thoughts are that the plan would work well enough. We already collect information from a good number of the young guys whose families get along with us, so arming them with weapons and posting them on the street corners wouldn't be that big of a step. There might be problems with the police, though. It will be hard for the Iraqi police to tell an armed terrorist from an armed security guard, and I think there's the possibility our guards might start to get a bit uppity with the police. The guns might go to their heads. We'll have to mitigate that."

Crider listened. "That's true. We'll also need to get them uniforms of some sort."

"We can get the uniforms made by one of the local tailors. It'll give him a good bit of business, and it'll tie the force to the community."

McDowell chimed in. "Fi-fi and Farquaad have a sewing shop, and they're always helping us out. Let's use them."

I nodded. "Yeah, that's true. They're good people, and they'll do fine with the uniform contract. Who's going to run the security force, though?"

Crider responded. "Captain Hamilton has a list of proposed names built from the people you and Rich have been working with. Cook has a similar list for his neighborhood. There won't be any leaders above the neighborhood level. Keeping the Sons of

Iraq at the local level is one of the concessions we made to the Iraqi government to avoid creating a parallel security organization in the country. As you might imagine, the Iraqis aren't thrilled about this whole idea, but even they recognize the necessity, and there's not much they can do to stop us anyway."

I soaked in everything Crider had just said. At my level, it all made sense. One of the problems we had was that even in a small neighborhood like *Abu Tayara,* we couldn't be everywhere at once. There were always corners of the neighborhood where insurgents could plant a quick IED when we weren't looking. Another problem was the number of unemployed, bored young men hanging around and causing trouble. By putting the young men in uniforms, posting them on street corners around the neighborhood, and using them to keep an eye on things when we weren't there, the Sons of Iraq could potentially solve both of our problems.

"I think it could work, sir. When do we start?"

"Captain Hamilton will keep you abreast of the schedule and let you know when we're ready to get started."

"Okay, sir. That sounds good. We'll be ready."

"I'm sure you will, Dan. Your guys have done a great job, and we appreciate the work you've been doing out here. None of what we're accomplishing would be possible without the basic blocking and tackling your platoon does every day."

"Thanks sir. These dirty bastards do all the work." I waved my hand at Van Awesome and McDowell. "I just walk around and drink Fanta all day."

Crider and Calahan chuckled. "Keep it up, then."

We talked for a few more minutes while Sergeant Major Jones and Kaluzny finished their conversation south of us. When they finished, Jones's truck rumbled back toward us, and Crider and Calahan shook hands with Van, McDowell, and me and climbed back into their trucks. Their convoy started its engines and left.

I turned to Van Awesome and McDowell. "Well, you just heard the way ahead from the horse's mouth. What do you think?"

Van replied. "Well, we're creating an armed gang in the *mahala.* I see the immediate value, but I wonder how it'll turn out long-term. At some point, we'll leave Iraq, and the Shi'a government is going to have to deal with the Sunni militias we create. I'm not sure how that'll turn out, but it probably won't be good."

I nodded in agreement. As usual, Van made a thoughtful point, but it was a long time before we realized just how prophetic it was.

18

Mahala 838, Iraq, November, 2007

M r. Barkhat's face looked as though it had been carved from stone as he steadfast-ly bore the wrath of the heavy-set middle-aged schoolmistress of the *Al-Najun* school. She was about a foot shorter than Mr. Barkhat, and her furious face, framed in a white *hijab,* poked out of the top of her black *jubbah* as she shouted at him. Her upraised finger hovered a few inches from his face, and her tone and demeanor were reminiscent of the friend of an aggrieved high school girl lecturing the guy who had just stood her up. My Arabic was still terrible, so I had no idea what the woman was shouting about, but it was clear she was upset about something.

The scene was fascinating as Ozzie, Gonzo, and I stood watching the pair from the entryway of the newly remodeled school.

"What's up, Ozzie?" I whispered.

As always, Ozzie was completely unperturbed by the wild outburst of emotion. "She says that when he built the school, he didn't provide enough desks for the students, and she wants him to bring more before the school opens."

"That's it? That entire outburst was over a few desks?"

"Yeah, that's it. She's not really that upset. That's just how my people communicate with one another."

"I'll never get used to it, Ozzie."

Mr. Barkhat was now responding. His voice was quieter but firm, and he had no accompanying hand gestures. His face was grave, and he didn't utter more than a few sentences. As he spoke, the woman crossed her arms and glared fiercely at him. When Mr. Barhkat was finished, he pulled out a cigarette, lit it, and quietly smoked. The woman continued to glare at him for another minute or so, then she turned to us. As she did so, her face went through a magical transformation.

Her brow unknitted, and her eyes lit up. Her scowl evaporated and was replaced by a broad, white-toothed smile, and her posture changed from one of cold standoffishness to one of warmth and welcome. She spread her hands wide and began to speak, pausing periodically for Ozzie to translate.

"Welcome to the *Al-Najun* school. I am the headmistress here, and I cannot thank you enough for what you have done to help us reopen the building. The children of the neighborhood have not had the opportunity to learn for many years, and we are very excited to start teaching them again."

I replied, again through Ozzie.

"It's our pleasure. We also want life to get back to normal as quickly as possible in the neighborhood and getting the kids back in school is a big part of that. Do you have everything you need to start classes?"

"Oh no, definitely not. We can start classes, but the children still need more desks, books, pencils, and supplies of all kinds. I have enough teachers, but it will be hard to teach the children well without these things. We must have more of them."

"I'll talk to the contracting office and see what they can do."

That had become my standard refrain when asked for something by any of our Iraqi counterparts. The Sons of Iraq needed more weapons. The school mistress needed more supplies. The generator operators needed more pay. The police needed more trucks. Every project we'd started seemed to need more of everything to keep running.

At first, I actually listened to peoples' specific wants and needs. Every time an Iraqi had an issue, we'd sit down over a glass of Fanta or tea, and I'd hear out his concern, carefully taking notes in my book. Since it was easy to find money to *start* a project, but much harder to find money to *adjust* an ongoing one, my goal in these discussions was always to try to help my aggrieved counterpart figure out a solution that didn't involve more money. This never went well, and many of the meetings devolved into scenes like the one I'd just witnessed between the schoolmistress and Mr. Barkhat. Nobody was interested in my suggestions, they just wanted to wheedle more cash out of the U.S. government.[1] I stopped having meaningful discussions about their problems. Instead, I fed them platitudes and trusted they'd figure out the solutions themselves using the resources we had already given them. While I felt this was an immature solution, it

---

1. Dr. Manza remained an exception to this rule throughout our deployment. I'm sure he made plenty of money off the U.S. government, but he always had the good taste to do the job well and to accept the profit he made without wrangling for more.

worked every time, and I decided that expediency frequently needed to trump principle.

The headmistress escorted us around the school to look at the work that had already been done. It was impressive. While the *Al-Najun* school hadn't been destroyed during the invasion, it had been in sad shape when we'd first arrived in the neighborhood. Many of the windows had been broken, most of the electrical fixtures and wiring had been plundered, and there had been extensive water damage. The rooms had been full of garbage and animal excrement, and the whole building had stunk of mold. Now, all of that was fixed. Mr. Barkhat's crew had even decorated the rooms with colorful cartoon characters and redone the landscaping. When we finished the tour, I turned to the headmistress and Mr. Barhkat.

"The school looks amazing. Mr. Barkhat, I will tell Dr. Manza that we are very satisfied with the work you have done. Headmistress, I very much like the way you have put the school together, and I look forward to seeing your students in class."

Both smiled and nodded to me as I turned and walked out of the school. As I was leaving, I heard the schoolmistress's tirade reignite behind us. Ignoring it, we left through the double-doored main entrance of the school and walked through the gates and onto Main Street.

Captain Hamilton's efforts to fund improvements in the *mahala* had been successful, and our mission today was to check on several of the current projects to make sure they were being completed on schedule. The school had been our first stop. Our next stop was one of the two generators we'd purchased to bridge the gaps in the Iraqi electrical infrastructure. It was located four blocks south of the school. We'd installed it on the site of the ramshackle Iraqi generator that had burned down spectacularly one night the previous summer.

It seemed as though everyone in the neighborhood was out and about conducting daily business, and as we walked south, we were greeted by dozens of the people we'd come to know over the past six months. Some of them, like Old Man Friend, a seventy-five-year-old man who'd served in the British Army as an Iraqi auxiliary, seemed genuinely happy to see us. Some seemed to appreciate that we'd waged a war last summer to restore their neighborhood so they could walk around it without fear of being killed. Others, like Haji Kramer, a tall Iraqi man with an uncanny resemblance to Michael Richards, greeted us more hesitantly since we'd arrested them in the past. Locals like Kramer, who had been found innocent by the justice system and returned home, were now in the clear from our point of view, and we had no issues with them. Still, for obvious reasons, our relationships with them rarely progressed beyond the exchange of strained pleasantries.

Awkward greetings aside, walking south down Airplane Street that day felt good. *Mahala* 838 was a small town–our small town-and we were the Andy Griffiths who kept it running smoothly. Being the sheriff of a growing small town felt a good bit better than being an occupying officer tasked with pacifying a war-torn neighborhood rife with sectarian violence. On the way to the generator, we passed Fi-Fi's shop, and Gonzo stopped in to say "hi." Fi-Fi was a French woman who lived in the *mahala* with her husband, Farquaad. The couple ran a store on Main Street that sold pretty much the same things all the other shops on the street sold vegetables, DVDs, canned goods, novelties, and cigarettes, but since Fi-Fi spoke excellent English and was friendly, we shopped there more often than in many of the other stores. Gonzo and 2nd Squad were particularly frequent customers.

Over the past six months, 2nd Squad had helped Fi-Fi straighten out several administrative issues with the local government, and, in turn, she had helped inform 2nd Squad about local goings-on. One of the quirks of counterinsurgency was that, since it required everyone in the platoon to interact with people in the neighborhood, relationships tended to develop haphazardly. This meant that as often as not, the relationships most critical to the platoon were not necessarily with the platoon leadership.

Ozzie and I waited outside while Schryver and Gonzo walked into the shop. They were in and out of the shop in just a few minutes. Gonzo bought a bag of fresh bread and a package of cream cheese for his guys to munch on and a bootlegged copy of *Hitman* to watch back at Falcon. He passed the supplies to his crew in the trucks, and we continued our walk.

When we arrived at the generator, we were greeted by two Sons of Iraq, who stood up from plastic lawn chairs to render us loose salutes. They were dressed in khaki uniforms and black combat boots, and they carried walkie talkies. We had scrounged them the radios from extras we had on Falcon, the boots had been procured by the supply guys from God-knows-where, and the uniforms had been sewn up by Fi-Fi and her husband on a contract we'd secured for them earlier in the month. Ozzie and I approached.

"Hi guys, how are things going today?"

The two Sons of Iraq looked at each other for a minute, then the shorter of the duo, Mahmood, responded. "Everything is very good, sir. We still need a replacement flare, though."

The flare needed replacement because of an incident the first night the Sons of Iraq had patrolled the *mahala*. Mahmood had been posted in the same place he was today, guarding the same generator. We'd given him a flare and told him to fire it into the air if he found himself in trouble so we could come help him out. About five minutes after we'd left, he'd fired the flare. We'd raced back to his position expecting trouble, but he was just standing there looking at us. When asked why he'd shot the flare, Mahmood said he'd wanted to test it to see if it worked. Flares only work once, so this was a problem. We had given him a second flare, ensured he understood not to fire it unless there was an emergency, and drove off. He'd fired it five minutes later and claimed the flare had malfunctioned.

I nodded, remembering all of this and thinking about how little I wanted Mahmood to have another flare. "I'll see what I can do."

We handed the two guards bottles of cold water from the trunk of Gonzo's truck and walked past them to the generator they were guarding. It was a huge Komatsu diesel generator, and it ran off a tank of fuel built into the building behind it. The generator was surrounded by a cage, and it had a corrugated metal roof over it. The power it produced was distributed by a wall of electronics similar to the cobbled together mess of circuit breakers that had burned along with the old generator, but the new one included insulated wire and had been professionally installed, so I didn't fear being electrocuted as I walked into the room. We ducked through the low cage door, and as we did so, the generator manager stood up and greeted us.

"Hello, sir. The generator is running very well, sir."

"That's good to hear. What percent of its overall capacity is it running at?"

"Not so much today, sir. The weather is good, so nobody is running their air conditioners. This keeps the drain on the generator low, which is good."

I nodded in what I hoped was a sage manner. "Great. That's good news." Despite having funded the generator's installation, I still knew absolutely nothing about managing a community's electrical grid, electrical capacity requirements, or really anything else about generators. To overcome this handicap, I'd asked Doc Osborne to give me a five-minute class on the subject before our patrol so I could ask some sort of intelligent-sounding question during the checkup. Nobody snickered during my conversation with the manager, so Doc's class had apparently worked well enough.

I shook the manager's hand and thanked him for his time, and we left to continue our inspection. Our next stop was a community center we'd sponsored at the north end of the *mahala*. This was around ten blocks from the generator, and a few months ago, when the weather was hotter and the streets more packed with hidden explosives, I would have mounted up and driven. Today, though, I decided to walk.

The walk took us zigzagging along the back streets of the neighborhood, past the local mosque and several residential streets. We passed Hamed's house, a man I knew who lived with his 126-year-old father. I'd been skeptical of his father's age when Hamed first told me, but Ozzie said his ID looked legitimate, and when I walked back and met the old man, his age seemed plausible.[2] Hamed's father was a shrunken, darkly tanned methuselah with a bald head and toothless mouth. He looked as old as any human I could imagine. When asked, he attributed his survival to a good lifestyle and regular prayer. Apparently, he prayed every time he walked through a doorway. When we walked past their house this morning, Hamed was in his yard washing the dust off his car and waved at us. We waved back and wished him and his father good health.

We turned north at the next street corner and passed General Abdullah's house. General Abdullah was sitting in a lawn chair enjoying the weather. I waved and greeted him as we walked by.

"Good afternoon, General. Have you remembered where the weapons of mass destruction are buried yet?"

Abdullah laughed. "Not yet, *mulazim*.[3] I'll let you know when I do."

During our initial census, Abdullah told us he was a general in the old regime. His stories and wall art seemed to back up his claim, so I'd asked him if he knew where Saddam had kept all his weapons of mass destruction. Abdullah had laughed and said they'd never had any. I'd laughed too, because if he *had* known where they were, I could have validated the whole war effort. After that meeting, the subject became a joke between us–an easy icebreaker with someone who could have been a real thorn in our side if he'd wanted to be.

At the end of Abdullah's street, we again turned east. We were nearing the community center when I saw Ginger hanging out with a group of his friends. He gave us a grin and a quick chin lift as we walked by, making an inquisitive gesture with his hands as though to ask whether we were going to handcuff him today. I shook my head, grinned

---

2. Even if the old man's age was calculated using the Iraqi lunar calendar, he'd still be 122 in solar years–a remarkable age in either case.

3. *Mulazim* is Arabic for Lieutenant.

back, and continued. Ginger was a freckled red-headed teenager–the only red-head in the neighborhood. We'd arrested him once for looking suspicious and taken him back to the COP for questioning, but it was quickly apparent that he was just a surly teenager, so we didn't take him to the DHA. Every time we saw him now, he asked if we were going to arrest him again.[4]

Finally, we made it to the community center, and on our approach I saw that McDowell's truck was already parked outside. The center was built into the bottom floor of a two-story strip mall. It had an external staircase and walkway, with storefronts lining the Main Street side of the building. The community center was on the bottom floor. I entered through the glass front door and immediately heard shouting to my right. I turned and saw a crowd of a dozen or so ten-to-fifteen-year-old boys mobbing a foosball table.

They were hooting and cheering at one of their number, who was skillfully manipulating the knobs on one side of the table. On the other side of the table stood McDowell. Despite being inhibited by his helmet, gloves, armor, and weapon, McDowell was vigorously spinning foosball knobs in an attempt to defeat the Iraqi teen at a child's game. The crowd loved it, particularly when the Iraqi kid beat McDowell about ten minutes later.

I looked around the rest of the community center. There were two other foosball tables, a ping-pong table, a TV, and a bunch of couches, chairs, and coffee tables covered in soccer magazines. Ozzie was across the room drinking a soda at an L-shaped countertop in the corner with a glass-front refrigerator next to it and a clerk selling drinks. To my eyes, everything looked exactly like it should.

The counterinsurgency field manual doesn't mention foosball, but in my view, the game McDowell played with that Iraqi kid probably did more to convince the next generation of Iraqis in our *mahala* that we weren't assholes than any of the military operations we conducted the whole year we served in 838. That foosball game may have even saved American lives. People don't blow up people they like, but conversely, you can't make people like you by blowing *them* up either. You can, however, *encourage* them to like you by playing foosball with them, and that's what we'd finally figured out in *Abu Tayara*.

The kids crowded around McDowell after his defeat to clap him on the back and hoot

---

4. We never did. We ended up hiring him as a Son of Iraq, eventually.

Arabic taunts at him. He pushed his way through them laughing and walked over to me.

"That little bastard was fast, but I could've beat him if this rifle hadn't gotten in my way."

I patted him on the back in consolation. "It's okay, McDowell. Even though it looked like the kid beat you, we actually won."

19

Mahala 838, Iraq, December, 2007

Fi-Fi's mother-in-law vigorously dug into the roasted lamb with both of her hands, ripping pieces of the carcass loose and depositing them on my plate. Her wrinkled face broke into a huge smile and her hands dripped hot grease as she gestured in a manner that expressed her hope that we enjoyed the meal. Fi-Fi's husband, Farquaad, spooned a mountain of aromatic, saffron-flavored rice on the remainder of my plate, and Fi-fi topped the whole thing off with a pile of dates and pushed the plate toward me across a table strewn with a wide variety of sweet and savory dishes, ranging from green sugar cookies and Jordanian almonds to pickled vegetables.

It was the week before Christmas, and I was seated at a large, round table in Fi-Fi and Farquaad's dining room. Doc Osborne and Gonzo were sitting on my right, and a blogger who Squadron had told us to take on patrol was sitting on my left with Ozzie. The room was richly decorated in traditional Iraqi style, with golden text from the Koran framed and hung on the walls, thick colorful rugs on the floor, and an embroidered gold tablecloth on the table. There was a beautiful chandelier overhead that bathed the room in warm light, and, incongruously, a Christmas tree in the corner. The Vienna Boys Choir sang "Oh Come all Ye Faithful" quietly in the background, and our armor and weapons were piled in a corner for a change–we wouldn't have been able to fit around the table otherwise.

Fi-Fi's family had set up the tree and prepared the meal as Christmas presents for our platoon. Although Fi-Fi had adopted her husband's religion when she moved to Iraq, as a French woman she still knew how to give her house a Christmas feel. Once she and Farquaad finished serving, Farquaad stood up and welcomed us to his home. His English was poor, so Fi-Fi translated for him.

"We would like to welcome you to our home this evening for a small celebration. In our country, it is the time of *Eid al-Adha*. In our culture, this is a time of great celebration, when we gather with our friends and family to serve the best meats and delicacies we have to offer in honor of Abraham's willingness to sacrifice his son Isaac when commanded to do so by Allah. We understand this is also a time of great religious significance in your country. It is good that our two countries can celebrate these times together.

"As I sit at this table with my own family tonight, I wish to express my appreciation for the sacrifice you are making for our neighborhood by being away from your wives and children for so long. The work you are doing here is wonderful. *Abu Tayara* is much better than it has been for the last few years. For the first time since the fall of Saddam, people are celebrating *Eid* openly in the streets without fear. For the first time since the fall of Saddam, our neighborhood has hope. For this, we thank you.

"Please sit with us tonight and enjoy our hospitality. Please relax and enjoy our expression of gratitude for what you have done."

After completing his speech, Farquaad sat down, picked up his fork, and began to eat. We all began to eat as well. Internally, I was glad Farquaad was happy with the progress we'd made in the neighborhood, but I was concerned he was being overly optimistic. Violence was down, but *Eid* was a perfect opportunity for an attack, and I was worried one might happen tonight.

The food was delicious. The blogger initially seemed a bit squeamish about the lamb, I suspect because of the way it was served, but social pressure and the delicious aroma of roasted meat eventually overcame his concern, and after a few minutes he was eating as enthusiastically as the rest of us. After washing down a bite of lamb with a swig of Fanta, the blogger piped up with a question.

"This is all excellent, and I appreciate the opportunity to join you tonight, Farquaad. I'm glad your family can be together to celebrate this occasion. The entire neighborhood really seems to be enjoying itself tonight. Was it always this way? What were the celebrations like before the invasion?"

Farquaad listened, then talked to Fi-Fi. After a few minutes of back and forth, she replied, "Before the invasion, *Eid* was always a time of great joy. It was our opportunity to travel back and forth between here and France to see both of our families and enjoy what our respective countries had to offer. The invasion changed that. We were scared to go outside at first, because the tanks often shot at people when they saw them. Then, after that, when there were no more tanks, we were scared to celebrate because the Shi'a militias murdered people or set off bombs whenever crowds gathered. Now, it is not like that though. Now, we can celebrate without fear."

The blogger nodded along, listening thoughtfully as he chewed on his food.

"Would you say it's better now than before the invasion?"

It was a sensitive question for a guest to ask over dinner, and I wondered how Farquaad would respond. I stared at the blogger pointedly to try to dissuade him from doing anything else socially awkward. There was a time and place for digging into the implications of the U.S. invasion, but now was not that time. He caught my eye and took the hint.

Farquaad's eventual reply was measured.

"I think Saddam was a bad person, and it is good that he is gone. He did many bad things to the people of Iraq. The invasion was also bad, though. It harmed many people and caused many problems. It may be that the bad things caused by the invasion were necessary to remove Saddam, but it would still be better if the bad things had not happened. Now, though, things are getting better, and we are happy. *Inshallah* this will continue, and things will be better from now on."

I think the blogger would have liked to ask more questions, but he seemed to understand that this wasn't the best time to do that, so he smiled, thanked Farquaad for his response, and continued eating.

The rest of the meal was delicious and relaxing. Gonzo and Fi-Fi gossiped about the goings on in the neighborhood. Farquaad looked satisfied with the way the evening was turning out and continued to offer food to his guests as we cleaned our plates. Ozzie chatted animatedly in Arabic with Farquaad's mother about who-knows-what.

After dinner, Fi-Fi brought out an ornate tray stacked with *istikanten* full of tea.[1] Iraqi tea is generally served with so much sugar that it's not so much drunk as chewed, so we chewed our tea along with the accompanying green, flower-shaped *Eid* cookies, then said our farewells and donned our armor and weapons to leave.

As we were getting ready to walk out, the blogger asked me a few questions. He was a middle-aged, bearded guy in glasses, khaki pants, and a reporter's vest. Bloggers and reporters were attached to us periodically, and they were generally pleasant enough. They would follow us around for a day, taking pictures and talking to people in the neighborhood, then go home and write a story about their experience. They always said that they wanted to see what the war was like, and I never had one act rude to me in

---

1. An *Istikan (istikanten* is the plural) is a small, clear tubular glass used to serve tea in Iraq.

person. Based on what they wrote once they were home, though, some clearly retained preconceived notions about what Iraq was like before they accompanied us.

"Lieutenant, that was an amazing experience. Do you usually get to do things like that on patrol, and do you get along that well with everyone in town?"

I laughed. "No, not so much. Three months ago, we were still getting blown up every day and finding dead bodies in the street every morning. It's gotten better, but we still spend a lot more time sweating in the streets than we do sitting down to dinner. As far as getting along with the people, we are on good terms with most of them, but there are definitely folks who don't like us. In some cases, that's because we arrested their relatives. In others, it's because we pissed them off by breaking something they owned. One lady is still mad at us because she thinks we blew up the gate to her house with a robot."

He nodded thoughtfully. "I've got to say, the war seems different than I imagined. I mean, in part, it looks exactly like I expected. The concrete barriers and checkpoints make the place look like a warzone. On the other hand, you guys don't do the sorts of things I imagined at all. I pictured you going on house-to-house searches, interrogating people, and searching cars at checkpoints. I figured this place would look like a police state, but instead, you mostly just seem to walk around talking to people."

"I'm glad it looks that way now. It took a long time and a lot of pain to get here, though. To tell you the truth, usually it doesn't even feel like combat anymore. Most of the time, I feel like a combination of Andy Griffith, a handyman, and a real estate agent. We solve petty neighborhood disputes, contract people to provide basic services, and interview people interested in moving into the neighborhood. We still arrest people sometimes, but we haven't had to raid a house in weeks. Instead, we generally just knock on the suspect's door, tell him we need to talk to him, and then take him to jail. When his family has questions, we tell them the truth, and explain that if he's innocent, he'll be home soon. If he's a bad guy, the family already knows it, and while they might not like us arresting him, they know they can't do anything about it. We haven't had any trouble arresting anyone for a long while."

The blogger listened as he pulled on his borrowed flak vest and helmet. "That's interesting. I think a lot of people back home would be surprised to see how much the people here appreciate your platoon being here. The war doesn't get much good press."

"That's unfortunate. I don't have time to watch the news, but we're definitely not killing babies, and I can assure you that I'm not getting any oil revenue off that refinery on the edge of town."

The blogger laughed as we finished strapping on our equipment and walked to the front door, where Schryver had been pulling guard during dinner.

"Sir, you're not going to believe what's going on in the street."

Schryver opened the door, and he was right. The scene that greeted us was unbelievable.

Night had fallen while we were in Fi-Fi's house, so as we stepped out onto the warm, damp Main Street air, *Abu Tayara* was bathed in the harsh white light of the newly functional streetlights. In their glow, we saw what looked like the entire population of the neighborhood enthusiastically dancing in the streets. Old men in *dishadashas* clapped to the exotic sounds of the *Oud* and the *Ney* blaring from '80s style boomboxes as young men in track suits hugged, shouted, and danced enthusiastically. Around the edges of the crowd, neighborhood women chatted with one another and clapped along to the music. And everywhere were gaggles of children. They ran between legs and perched on every wall and gate I could see. They waved sparklers and threw firecrackers. They cartwheeled, somersaulted, and chicken-fought with one another, and above all, they begged for candy.

As I pushed out into the crowd, a thousand tiny hands stuck out and asked for a treat.

"Mistah! Mistah! Candy! Candy!"

The limited amount of candy that Ozzie, Gonzo, Schryver, and I carried in our pockets was exhausted almost immediately. Despite our protestations, the kids continued to follow us through the crowd, steadfastly refusing to believe that a country with the wherewithal to project military power six-hundred and ninety-eight thousand miles away from its shores would have failed to equip its soldiers with sufficient Tootsie Rolls to appease the *mahala's* children. Eventually, though, they realized we weren't just playing hard to get. Most of them vanished into the crowd, but a few hung around on the off chance they would get lucky after all. One stood in front of me, holding up his arms and grinning broadly. He shouted something unintelligible over the background noise of the *Eid* celebration. Ozzie leaned over and shouted into my ear.

"He wants you to put him on your shoulders."

I shrugged. I could do that. I reached down to grab the kid under the arms, spun him around, and sat him on my shoulders. I was still wearing a helmet, armor, and a slung rifle, so the process was a bit awkward, but once I got him situated, he fit well enough. The crowd immediately parted around me, and I found myself in a ring of clapping, pointing Iraqis. Flashes went off everywhere as people snapped pictures of what was an undoubtedly ridiculous looking situation. Ozzie looked around laughing, cupped his hands and shouted, "Hey LT, they want you to dance."

This wasn't what I planned on, but I couldn't think of anything else to do in the situation, so I danced.

In the best of circumstances, I'm an awful dancer. With an Iraqi kid on my head and 65 pounds of kit on my back, I probably set a new low mark. The only saving grace was that in 2007, footage of something so ridiculous couldn't go viral, so only still photos of the incident ever emerged.

The Iraqis were delighted, and they continued to hoot and applaud until I put the kid down and took my bows. I turned to Ozzie, Gonzo, Schryver, and the blogger, waited until their laughter subsided, and then waved them out of the crowd toward the trucks, which were all parked on the relatively unobstructed side streets to watch for trouble. The first trucks we came to were 3rd Squad's, and we found that Van Awesome and McDowell had also gotten into the spirit of *Eid*.

A local Iraqi man had dressed himself up in a furry orange bear costume to advertise his business. He hadn't been able to find a head for the costume though, so he'd hollowed out a two-foot-tall orange stuffed bear, cut a hole in its stomach, and stuffed it on his own head in place of a hat. The name of the business was scrawled in Arabic across the stomach of his suit in black marker. To help his cause, Van Awesome and McDowell had hoisted the man onto their shoulders and were carrying him around the neighborhood as he waved and shouted. I took a picture of the trio for our daily report and walked over to talk to them.

"That is quite a costume. Is this how you've decided to bond with the people of the community?"

"Just doing our part for the war effort, sir. We can't sell war bonds, so carrying this orange bear guy around seemed like a reasonable alternative."

"That makes sense. I think you're spot-on. Carry on, men. Literally."

I left them to their community engagement and walked off with Gonzo, Schryver, Ozzie, and the blogger to find 2nd Squad. Gonzo called his guys on the radio and discovered that his trucks were north of us, so we walked along a side road to find them without having to negotiate the *Eid* crowd.

As we made our way along the dark, quiet street, the sound of fireworks and shouting echoed down from the festival. In the distance, we saw a blaze of flames appear on the ground as a group of Iraqis used lighter fluid to create a wall of fire in the street and take turns leaping through it.

The blogger turned to me.

"Yeah, this is nothing like I imagined it would be."

"When Sherman said war is hell, I'm pretty sure this isn't what he meant. Still, don't let the party fool you. There's still a fair chance someone will detonate a bomb in that crowd tonight and roll back all the gains we've made here. Also, the unfortunate reality of counterinsurgency is that if things have cooled down here because we've succeeded in stabilizing things, someone else is probably having a rough time not too far away. We've built a system that makes it difficult for insurgents to hide in *Abu Tayara*, but we didn't arrest or kill all the bad guys, and the ones we didn't deal with have probably just left our neighborhood to find one where it's easier to hide. Unless we can apply pressure evenly across the entire warzone, the insurgents will exploit any disparities they can find and drift toward the less secure areas to continue causing problems. 1-28 is working in a denser urban area in West Rashid, just a few kilometers west of here, and they're having a hell of a time securing the place. In part, that difficulty is probably driven by our success here."

The blogger nodded and listened. Gonzo grabbed his arm to prevent him from stepping in a puddle.

"Watch that puddle, sir. I stepped in what looked like a shallow puddle a few months ago and sank up to my thigh in sewage."

The blogger thanked him and stepped around the puddle as Ozzie laughed loudly and clapped Gonzo on the shoulder while recounting his earlier misfortune. After that incident, one leg of Gonzo's pants had been dark brown from boot to crotch. Since ACUs stained easily, he had worn a pair of pants with a single brown leg for the rest of the deployment.

The blogger was silent the rest of the way to the trucks. I looked over at him as we walked and wondered what sort of story he was piecing together in his head. From my guy-on-the-ground point of view, bloggers and reporters were a necessary evil. I understood the need to message what we were doing in Iraq to the people back home, because it was important to counter the narrative that we were all soulless baby-killers or poor corporate stooges; taking a blogger with us on patrol was a great way to show the world firsthand what we were actually doing. On the other hand, having bloggers on patrol could be risky. My guys were soldiers who'd been working closely with the Iraqis under dangerous conditions for months. They sometimes said or did things that could look pretty bad when taken out of context, and some of the bloggers we worked with were clearly interested in viewing things through whatever lens underscored the message they wanted to carry home. A bad media story could set the entire mission back. What sort of message, I wondered, would this blogger carry home?

When we reached the vehicles, I told Gonzo to keep the blogger with him for the rest of the evening. The blogger seemed to have seen all he wanted to see, and he looked tired from walking around in armor for the last few hours. Besides, we still had a few hours until we were relieved, and I wanted to see how the Sons of Iraq were handling the *Eid* crowd. I wasn't sure how the guards would appear, what they might say, or what the squadron had already said about the Sons of Iraq, so this wasn't anything I wanted to discuss.

"Come on, Schryver, let's see how our militia is doing this evening."

As we walked west toward Main Street, the sounds of the festival grew back into the roar of shouting, music, and fireworks we'd heard outside Fi-Fi's house. I smiled and took a moment to enjoy the relative sanity of a wild street festival. It sure beat cleaning dead bodies out of a burning truck or getting blown up by a booby-trapped building. It looked like we might make it through the festival of *Eid* without an attack. "Maybe our approach here is working," I thought. "Maybe we'll stabilize this place after all."

20

Mahala 838, Iraq, January, 2008

"I don't care what his fucking religion says about the situation. We're taking his wife to the hospital."

As Ozzie translated my response, the man's face reddened. I couldn't tell if he was actually mad or just pretending to be mad to save face, but at that moment, I didn't care. The woman was going to the hospital.

We'd been busy preparing for General Petraeus's arrival later in the day, but our work had been interrupted when a woman in a pink headscarf had flagged us down to tell us her sister was sick and needed medical care. This wasn't an unusual request, and Doc had stitched up cuts, diagnosed illnesses, and given medical advice to dozens of neighborhood people. So Doc Osborne, Ozzie, and I followed her to her house to see if we could help. We parked outside, and the woman escorted us in. She led us into a living room, where a heavyset middle-aged man was sitting on a long, low, red couch. He introduced himself as the sick woman's husband. The woman who'd escorted us in offered us Fanta and a cookie from a silver tray, then sat down and discussed her sister's symptoms with Doc. After a few minutes, Doc turned to me.

"Sounds like she might be having a miscarriage, sir. I'll have to take a look at her to be sure though."

"Okay, makes sense. Go for it, but make it quick. We're a little tight on time."

Ozzie translated the request, and the man immediately stood up, crossed his arms, and said something curtly in Arabic.

Ozzie turned back to me. "He says we can't examine his wife."

"Why not?"

"Doc isn't related to her. He says unrelated men can't touch his wife."

"Is that a thing, Ozzie?"

Ozzie shrugged. "Yeah, some people are like that. It's a religious thing for some Muslims."

I turned to Doc. "What happens if she doesn't get treatment?"

Doc shrugged. "It's hard to say. Maybe she'll be alright, or maybe she'll develop sepsis and die. Without taking a look at her, I can't give you a very good estimate, but if she's got a fever that is as high as her sister is saying it is, she's probably in trouble."

I turned back to Ozze. "Tell her husband that. Tell him his wife is going to die if we don't get her medical treatment."

Ozzie did so, and the man's face hardened. He shook his head and unleashed a flurry of Arabic, accentuating his speech with exaggerated hand gestures. I didn't wait for Ozzie to translate.

"I don't care what his fucking religion says. We're taking her to the hospital. If he has a problem with it, he can take it up with the Iraqi police."

The man was clearly fuming as I coordinated with Kaluzny on the radio to get the woman to the hospital in the Green Zone. I was still an armored foreigner with a gun, though, so whatever he was thinking didn't amount to anything more than a dirty look as Doc left the room with the sick woman's sister to prepare her for the trip.

"Ozzie, tell her sister she can come as an escort."

The preparation didn't take long, and Kaluzny pulled up outside just as the two women were ready to go. Kaluzny walked in to help Doc load up the two women, giving the woman's husband a withering look as he moved through the house. Iraqis were generally intimidated by Kaluzny's enormous size and severe looking visage, and the disgruntled husband was no exception. As we left the house, I told Ozzie to inform the man we'd have his wife back by the next day. The man didn't respond, nor did he say anything as we pulled away from his house, Kaluzny to head to the Green Zone and me to link up with Blanco's Squad to continue preparing for Petraeus's impending visit.

Since December, things in the neighborhood had run smoothly, and most of our efforts were showing considerable progress. As well as things were going in the *mahala* though, and despite both our efforts to understand the local customs and the local people's desire to avoid running afoul of us and getting arrested, we still had occasional friction. Like today's issue, the friction mostly arose from foundational differences between American and Iraqi culture. Today, that difference had been a conflict between whether a woman's right to healthcare trumped her husband's right to adhere to the tenets of his religion. Another day, it had been our opposition to the Iraqi tendency to keep mentally disabled people chained and collared in the family basement. A third day, it had been our attempt to unravel the tradition that allowed the nomadic bands of trash ninjas and their flocks to roam the neighborhood at will.[1] In each case, we'd had to decide whether stepping on cultural tradition was necessary to avert what we saw as a greater problem, and it seemed like whatever we decided, the result was always friction.

I checked my watch as I climbed into Blanco's truck. I only had a few hours until Petraeus showed up, and I needed to complete the last preparations before the circus that accompanied him arrived. Our success in the *mahala* had been double-edged. On the one hand, we suffered almost no attacks anymore, crime was down, and the neighborhood had become a shining example of success for the U.S. war effort. On the other hand, all the positive news had brought a stream of distinguished visitors. Admiral Mullen, chairman of the Joint Chiefs, Ambassador Crocker, the ambassador to Iraq, and Major General Odierno, the overall commander in Baghdad, had each visited the *mahala* to see how well things were going, to shower praise on our unit, and to have a meal with Dr. Manza, the man we credited with much of our success. Unfortunately, hosting these visitors significantly disrupted our normal work in the *mahala*, as each visitor brought a huge logistic and security package that we had to accommodate, and before each visit, every commander between us and the visitor wanted a pregame brief with us to make sure we weren't going to say anything too controversial.

Given that today's visitor was the legendary David Petraeus, the overall commander of the war and the creator of The Surge, the pregame workload for his visit had been substantial. Over the last week, every aspect had been briefed, re-briefed, dissected, planned, and rehearsed, and today we were working on the final on-ground details. First, we had met with Dr. Manza to discuss the visit and ensure he'd have a full dinner prepared for Petraeus, his staff, and an assortment of distinguished invitees from around the neighborhood. Then, we'd met with the Iraqi police to make sure the police were ready to provide escorts for Petraeus's convoy and that they were ready to close the

---

1. We thought they had ties to terrorist organizations, but as it turns out, they didn't. As a Texan, the idea that a group of people could entirely ignore all property rights and graze their dirty herds of spray-painted sheep wherever they liked was anathema. Our interactions with the trash ninjas were further inhibited by a language barrier; even our interpreters often couldn't speak to them.

entrances to the neighborhood while Petraeus was on the ground. Finally, just before the incident with the miscarrying woman, we'd met with the headmistress of the *Al-Najun* school to make sure she was ready to answer any questions Petraeus might ask if he decided to talk to her. The only thing left before we went to wait at COP Banshee for the helicopters to arrive was to ensure the Sons of Iraq checkpoints were ready for the visit. With the criticality of the Sons or Iraq to the overall war effort, I was sure Petraeus would want to engage with them if he saw one of their checkpoints.

Blanco and I drove to the northernmost Sons of Iraq checkpoint, near the northern generator, to talk to the team there. When we arrived, Mustafah, one of our tallest guards, was sitting in a lawn chair. He stood and greeted us as we walked up, then turned immediately to Ozzie and started to speak. Ozzie listened and translated.

"He says he's been having trouble with the police again. Yesterday, they stopped his family and harassed them as they were coming through the northern gate. He says the police don't like the security force and take every opportunity they can to harass its members."

These complaints weren't new. Since we'd started the Sons of Iraq, there had been friction with the police. As Americans, our long-term agreement with the police to patrol the *mahala* in their stead worked fine, but the police seemed much less amenable to armed Sunnis walking the streets.

"Ask him if the police actually *did* anything to his family."

Ozzie listened to Mustafah's response and turned to me with a shrug.

"It sounds like maybe they said rude shit to his family or threatened them or something. Nothing concrete. Maybe bullshit, I think."

I nodded. "Tell him I'll talk to the police for him and see if I can get them to stop. Also, tell him to make sure he and the rest of the Sons aren't provoking the police, either. I get similar complaints from the police at the checkpoints about Sons of Iraq threatening them with weapons, and I can't get the police to calm down if the security force is also causing problems."

Ozzie translated, and Mustafah rolled his eyes. He and the other Sons of Iraq always acted petulant when the Iraqi police were involved. When Ozzie finished, I continued.

"Also, ask him if he's ready for the big visit today. The guy coming is the one who signs all the checks for everything we're doing here, and if the Sons of Iraq fuck up the visit by bitching about trouble with the police, I'll be pissed off."

Ozzie relayed some version of my message to Mustafah, and given that he took it well,

I suspected Ozzie had softened it somewhat. That was fine. As long as everything worked out, I had no problems with Ozzie tweaking my comments to better suit the audience.

I shook Mustafah's hand, and as we left, he picked up his radio to relay our conversation to the guards at the other checkpoints. I hated being a mediator between the Sons of Iraq and the Iraqi Police. When we'd created the security force, I'd imagined that at some point the Sons of Iraq would become some sort of Sunni arm of the Iraqi police force. But if they couldn't even speak civilly to one another, how were they ever going to merge into one organization? Our mediation allowed the *mahala* to function, but the friction between the two security organizations indicated there would be long-term problems that I wasn't sure how to solve.

As we made our way to the other checkpoints, I noted the continued improvement of the neighborhood. The curbs had all been repaired and painted, the ugly, gray T-Walls surrounding the neighborhood were now painted with colorful scenes of Iraqi life, and the streets were generally clean. Almost all of this had been arranged to some degree by one of Dr. Manza's friends or relatives. After working with other Iraqi companies had resulted in blown deadlines and poor results, the command decided to run almost all of our reconstruction efforts through Dr. Manza. At our insistence, he subcontracted work to other people in town, but only people he trusted to do the work well, and he kept a hand in every project. If Dr. Manza had been killed, most of our programs would have immediately collapsed, but while we recognized the fragility of this approach, *we needed results*, so we traded fragility for effectiveness.

Even the trash collection and propane distribution were working well, although again, the services were subcontracted through one of Dr. Manza's relatives rather than the Iraqi government. The local municipal director, a Shi'a, had told us 838 was too dangerous for his people to service, so we'd created yet another work-around. It was these work-arounds that were starting to bother me. If everything we built was a workaround to the actual system, what would happen when we removed ourselves from the equation? The friction I was seeing between the Sons of Iraq and the police, and the unwillingness of the government to provide services to the *mahala* without our intervention, both suggested that the aftermath of our pullout wouldn't go well. As there was no pullout scheduled for the moment, though, solving these problems could wait. The next unit rotating into sector could figure it all out.

While we were checking the last Sons of Iraq position, I heard the distant sound of UH-60 engines. The roar grew in volume as the helicopters approached. I checked my watch. This had to be Petraeus landing at the COP. I climbed back into Blanco's truck and put on my headset. The company net was alive with traffic from Hamilton, Whitmore, and X-ray as they talked through the details of the command visit. Petraeus was at the COP now. He'd link up with our company, squadron, and brigade commanders there, and they'd all move into 838 by truck, disembark their vehicles at the southern end of the neighborhood, and walk north through the *mahala* to Dr. Manza's house. Our platoon's role in this plan was to provide our normal presence in the neighborhood while the command visit was in sector. My role was to link up with Hamilton to walk with the command teams and answer any questions that came up. Everything in the neighborhood was ready. All Blanco and I had to do was drive to the southern checkpoint and wait.

We didn't need to wait long. Within ten minutes, the first trucks from Petraeus's security element drove through the checkpoint. Around the tenth truck, I realized I had wildly underestimated how many people were involved in this visit. The security element was *huge*. As the vehicles continued arriving, they began disgorging security personnel, and I quickly lost count of the number of Humvees, MRAPs and dismounted guards present in our small neighborhood. The ground guys fanned out expertly, climbing onto roofs and running north to post themselves at every street intersection for several blocks. The trucks spread out too, blocking traffic and keeping the street clear of local people. Blanco and I just stood and watched, marveling at the scale of it all.

When security was set, the leadership vehicles started to arrive. Hamilton arrived early and walked over to stand with me. Rich was with him. His platoon was on COP security that day, but he'd decided to take part in the command visit to see how it went. Hamilton, stone faced, watched the spectacle unfold in front of us.

"What do you think, Dan?"

"Holy shit, sir. That's way more people than I thought would be involved."

"Just wait. You haven't seen the collection of strap hangers that came with the big man."

Hamilton was right. As the leadership vehicles arrived, they too disgorged dozens of people. Field grade officers and senior NCOs of all ranks prowled the streets; colonels, lieutenant colonels, majors, and sergeants major were everywhere. I hadn't seen this much brass in one place in my entire career, much less in our tiny slice of Baghdad. Aside from the preponderance of rank, the main thing I noticed about the group was its cleanliness. The Army's current field uniform–the grayish white ACU–was notorious for showing stains, so my platoon's armor, helmets, and uniforms were almost brown from the months of sweat, grime and worse that were ground into the fabric. Not so the fresh bunch of Green Zonians who flooded 838 today. Their uniforms were crisp and clean, their armor was spotless, and their unused weapons were pristine as they milled about, chatting with one another. For many of them, this may have been their first time out of the Green Zone, and their equipment reflected it, with the pouches and accessories on their vests arrayed in a way that no soldier who lived in sector would ever array them. No doubt, many of them hoped something dangerous would happen today so they could earn a quick combat decoration.

In the middle of all of this, Petraeus himself stepped out of his vehicle. He was unarmored and unhelmeted. When you're the HMFIC you can do that sort of thing. He looked unlike everyone around him.[2] Petraeus was short, thin, and slightly stooped, and he wore a simple field uniform that looked clean but well-worn. His face was old and lined, and he would have looked entirely unremarkable except for his expression. In Army parlance, people who look dynamic, intense, and in-the-know are said to be "switched on," and despite his unassuming physical demeanor, Petraeus came across as being absolutely switched on. It was his eyes that conveyed that impression. They were completely alive and seemed to miss nothing, and they projected more intensity than I've ever seen in anyone else.

As Petraeus stepped out of the truck, he was quickly surrounded by concentric rings of attendees. Our Brigade Commander, Colonel Gibbs, and Lieutenant Colonel Crider were in the innermost circle; as the most senior commanders familiar with the sector, they occupied a privileged position in the upcoming procession. The next ring comprised the highest-ranking individuals present, the generals, colonels and lieutenant colonels who made up Petraeus's senior staff. Beyond them, the scrum got more loosely hierarchical and less organized. The mishmash of personal security guys, bomb dogs, aides-de-camp, executive officers, and public affairs representatives formed a sort of amoeba around the nucleus of important people. Then came the formal cordon of security guys who ensured no unauthorized personnel interrupted the party, and past *even them* was me, Blanco, Smith, and Hamilton, standing off to the side and gaping at the spectacle. Hamilton sighed.

---

2. The HMFIC is Army slang for the "Highest Mother Fucker In Charge." It's a situational title, that's more informally understood than actually awarded. Nobody in *mahala* 838 that day had any question that Petraeus was the HMFIC.

"I guess I should join the head shed."

"That's all you, sir. I think we'll stay out here. You guys are probably the biggest VBIED target in the hemisphere, and someone will need to stay alive to call the MEDEVAC after you get blown up."

Hamilton acknowledged the joke with a nod and pushed his way in to join Crider and Gibbs. The ACU-patterned amoeba began sliding north along Main Street, extending pseudopods of security personnel and photographers as it went, and retracting them as the nucleus passed. Rich, Blanco, and I ambled along a few blocks behind the whole thing, still marveling.

"What do you think, Rich? What the fuck is going on right now?"

Rich shrugged. "Brother, I have no idea. What a zoo."

"You think Petraeus wants to hear anything from us?"

"Nope. I don't think anyone's interested in that at all."

I nodded. Rich was probably right. What did we know, really? Probably nothing that was applicable at Petraeus's level. Plus, anything we said would be subjected to scrutiny by the dozen levels of command between us and him, which would result in paperwork, briefings, and probable ass-chewings, all of which would cut into chow time.

"Yeah, you're right. I feel bad for Dr. Manza. Can you imagine that mass of humanity showing up in your dining room?"

"Mandy [Rich's wife] would flip out. Still, he's making plenty of money working with us. I'm sure dinner won't set him back too much."

"Fair point."

As we walked north, Petraeus stopped to chat with a few key locals we'd prepped for the occasion, like the headmistress and the Sons of Iraq. I hoped they weren't saying anything that was going to cause me extra work. It was while he was talking to Mustafah at the northern generator that I noticed something weird about the scene. There were almost no Iraqis in Petraeus's party. There was one army commander, a police chief, and someone who might have been a government official, but nobody in the Iraqi system that was comparable to Petraeus. Even the interpreters he'd brought were American. My brain turned this fact over as we continued north, and when the nucleus arrived at Dr. Manza's house, and I saw General Petraeus shake the Doctor's hand and thank him for the work he had done to rebuild the neighborhood, I realized what was bothering me.

The commanders in the Green Zone were making the same mistake at the national level that I was making in 838. Just like we were building workarounds for the systems that didn't work in our neighborhood, they were building workarounds at the national level. *The entirety of the security and reconstruction apparatus in Iraq was dependent on American forces and American money.* Our effort was focused on building influence between ourselves and the people of Iraq, rather than between the government of Iraq and the Iraqi people.

"Oh shit," I thought. "How is this going to end well?"

Part V

Closing Time

21

Mahala 838, Iraq, April, 2008

I n *The Hobbit,* J.R.R. Tolkien wrote, "Now it is a strange thing, but things that are good to have and days that are good to spend are soon told about, and not much to listen to; while things that are uncomfortable, palpitating, and even gruesome, may make a good tale, and take a deal of telling anyway." Similarly, while not necessarily "good to spend," our last five months in 838 were the sort of days that aren't much to listen to.

There were funny incidents, like the week we spent filling Tigernet with photos of all the stray animals in the neighborhood–we called it the 838 Safari. Or our month-long effort to stack the entire Comanche Troop section of the brigade 'yearbook' with nothing but photos of McDowell. There was the day when we tried to stage a bigfoot sighting. And once, at a meeting, the Iraqi police commander tried to sweet talk me into ordering a tattoo gun from the States for him.[1] There was even a weeks-long saga during which I tried to convince Captain Hamilton to let our platoon conduct a cultural immersion patrol to the ruins of Babylon.[2]

None of these make for much of a story, though. Our last five months mostly consisted of long, quiet days spent checking up on our security forces, dealing with small town disputes, keeping the neighborhood running smoothly, and trying our best to integrate what we'd built into the Iraqi government's systems.

We never made much headway on this last effort. Every time we thought we were

---

1. I told him I'd order him one if he got me one of the famous knives carried by Saddam's *Fedajin.* Neither of us came through with our end of the deal.

2. It took me several tries to even convince him I was serious. Once he believed me, he turned me down cold. In hindsight, it was probably the right call.

getting somewhere, we hit a hurdle that halted our progress. In the spring, for example, there had been a formal, national-level push to integrate the Sons of Iraq into the Iraqi police. We'd brought the security guys to COP Banshee, given them all physical exams, and worked to standardize their equipment and uniforms. The effort had gone smoothly enough until the government had demanded full personal and family profiles of our guys. I don't know why we didn't foresee this, since they were asking for the sort of information new recruits would normally provide, but when the order came down, we balked. What would the government do with the names and addresses of every armed, young Sunni in the *mahala*? It didn't take much imagination to figure that one out. Once we left, our Sons of Iraq would be arrested and killed.

Our efforts to integrate utilities and services in our neighborhood into the Iraqi systems were similarly futile. Hamilton and Crider had dozens of meetings with the regional leaders at the *belladia*, but none of the meetings went anywhere. The regional government was clearly uninterested in providing services to Sunnis. Despite all our efforts, Iraqi resistance to integrating Sunni and Shi'a defeated our attempts to force the issue, and we accepted the American-enforced status quo we'd built in the sector. We weren't getting killed anymore, and we saw daily evidence of the positive contributions we'd made to the *mahala,* so acceptance came easily. Our spring days blurred into a hot, unremarkable series of long, quiet patrols.

My calls with Alycia had become similarly unremarkable. After R&R, I'd fallen back to my old habits and focused too much on the war. Even when things slowed down in sector, I neglected to invest enough time calling or writing home, and our conversations had settled into a series of dull, sad phone calls and email exchanges. Nothing terrible happened, but it had been so long since we spent any real time together that we just didn't have much left to talk about, and there didn't seem to be much emotion behind any of what we did discuss. Neither of us was happy with the situation, but there didn't seem to be much we could do about it.

Nothing significant changed in the *mahala* until mid-May, when we heard rumors that our relief was due to arrive soon. This news was initially greeted with skepticism by the boys; after the last extension, there wasn't much faith that our war would end anytime in the near future. However, once the initial rumors were followed by advance parties, published relief timelines, and requests for information about how much equipment we'd need to ship to Kuwait, the platoon started to get excited. Finally, the day arrived when I met the replacement platoon leader who'd be taking over 838 after we left.

The New Guy looked like all new guys. His uniform was clean and crisp, his equipment was in immaculate shape, and his demeaner combined the false bravado and wide-eyed uncertainty that we all have when we're trying to look cool but have no idea what's going on. We met outside the barracks at FOB Falcon, shook hands, loaded up into McDowell's truck, and rolled into sector.

"Hey man, how was the trip over?"

"It went smoothly. No real problems. The bulk of the unit is at Camp Victory, and they'll be deploying to Falcon next week."

"Okay, how long do you have to RIP with me?"

"Just today, man."

I laughed. "Just today? Are you kidding? I can't even introduce you to all the people you'll need to meet in a day, and we certainly won't have time to discuss the contracts in progress or the construction efforts we're working. Why are you getting so little RIP time?"

"Your brigade is getting replaced by our battalion, so my company will be covering down on everything your squadron has now. I've got to RIP with you today, A Co tomorrow, and B Co the next day. I head back to Victory this weekend."

And there it was.

We'd been so successful at fixing the situation in *Doura* that our successors had determined they didn't need to devote any resources to ensuring the continuation of our success. In a way, the decision made sense. Military planning teaches us to assess the greatest threat and devote the bulk of our resources toward defeating that threat. Our replacement unit would need to secure a much larger area than we'd been responsible for because the additional forces that had been pushed into Iraq for The Surge were leaving, so our successors had justifiably concluded that since 838 and 840 were largely pacified, they could devote more forces to West Rashid, which was still very violent.

As reasonable as this decision was, it overestimated how durable the stability we'd created in 838 would be when we were no longer there. Without U.S. forces to maintain the peace between the Sunni and Shi'a, it was only a matter of time before the security situation unraveled.

Force allocation issues weren't addressable at the platoon level, though, so I turned my attention back to the RIP.

"Okay. I'm not sure that's going to work out very well, long term, but we'll do what we can with what we've got."

As we passed various landmarks on our way into sector, I pointed them out to The New Guy.

"All that on the right is Mechanics. It used to be part of our sector, but now some other unit owns it. We get shot at from Mechanics sometimes, but otherwise, we've got no business there."

"I think we'll be owning that, too."

"Shit. Yeah, it's not a great area. Good luck with that."

A few seconds later, I pointed again. "That's the *belladia*. It's a sort of combination municipal court, utility company, and city hall. Our company and squadron commanders go there periodically to wrangle with the Iraqis about supporting our neighborhood, but the people who run it are Shi'a, so we've never had much luck working with them. The only time our platoon goes there is to escort the propane tractors on the rare chance the *belladia* coughs them up. We don't have time to go today, though, so you'll have to meet them on your own."

"Why do you care about propane?"

His question momentarily stunned me. What did he mean? How the hell could anyone not know why we needed propane? Was this guy an idiot? Where had he been, that he didn't know about something as basic as *the criticality of propane to counterinsurgency*? Then it hit me: I was the idiot. The New Guy had been in Colorado, or Kentucky, or wherever his unit was based, and he'd spent the last year running his platoon through the same sorts of training events we'd run through back at Riley, none of which involved propane. It had taken me a year in Iraq to understand the economic and political underpinnings of *Doura,* and this guy had been here exactly one day. I was talking entirely over The New Guy's head. I needed to slow down.

But I couldn't. I had an overwhelming amount of knowledge to impart to him, and I only had a few hours in which to do it. As we rolled toward the *mahala,* I found myself at a loss for words. I couldn't think of anything useful to tell The New Guy which didn't require so much explanation that I couldn't make it make sense in the time we had available. How could I explain the friction between Sunni and Shi'a in a few hours? How could I convey the nature of the triangular relationship between us, the Sons of Iraq, and the Iraqi police in a single afternoon? How could I explain our relationship with Dr. Manza or summarize the dozens, if not hundreds, of meetings we'd had with him about how to keep the *mahala* running?

I couldn't, of course. That was the problem. If I'd had a week of The New Guy's undivided attention, I could have covered a tenth of what I needed to explain, but without the context which only experience could provide, he would have only retained a quarter of even that small amount of information. By my hasty math that meant that after this RIP, The New Guy would have only about .3% of the information he needed to carry on our work here once we left.

.3% is a pathetically, depressingly small number, but .3% was probably as much of our hard-won knowledge as I'd be able to impart.

As we pulled into the northern checkpoint, I stopped and introduced The New Guy to the Iraqi police. The three of us exchanged pleasantries, shook hands, and spent ten minutes talking through the current security situation in the neighborhood. On the way out of the checkpoint, The New Guy and I climbed back into our truck, and I tried to explain our relationship with the police.

"Alright, man, so when you arrest someone, you'll need two sworn statements from witnesses who can attest to the arrestee's crime. If the bad guy is Shi'a, you probably want to take him to the DHA, because the police will just release him. If he's Sunni, you need to determine whether he's *really* bad, or just *slightly* bad. *Slightly* bad Sunni criminals should also go to the DHA, although afterward you'll want to swing by the arrested guy's house to explain to his family why you picked him, or you'll piss off the locals. *Really* bad guys can get turned over to the Iraqi police, but once you do that, you can't be sure the police won't murder them, so we don't do that often."

The New Guy looked at me, nodding. He jotted down a few notes in his waterproof notepad.

"Did you get all that? Does that make sense?"

He said he got it, but I could tell that he probably didn't. Fuck. There was just no way to make this RIP work. The New Guy interrupted my internal monologue with a question.

"Have you guys hit many IEDs?"

"Yeah, a bunch, but none lately."

"Oh."

"But do you get how the detention process works? It's really important."

"Yeah, I think I got it."

That's how the rest of our day went. I tried to distill hideously complicated problems into sixty-second blurbs and then provide our recommended solutions. The New Guy scribbled a few notes, nodded, and then asked a question about something *he* thought was critically important, like MEDEVAC procedures, calling for fire, or coordinating with aviation. That information was important too, but his questions were about such elemental tasks that after a year of combat, they'd been so automated by the platoon, that sometimes I couldn't even remember the answers.

With a heavy heart, I finished the RIP. We dropped The New Guy off with Apache Troop so he could talk to them about 840, the *mahala* next door, and went about our business. We didn't see our replacements again until they arrived at Falcon a few weeks later and started moving into the barracks buildings. The rest of the relief process was unremarkable, and both of our units worked through the administrative details of replacing one another without incident.

Our normal schedule continued until 1st Platoon was due to conduct its last patrol in *Mahala* 838, a surprisingly bittersweet experience. On the one hand, I was ready to go home. I was sick of the hot dusty days, sick of dealing with the petty squabbles, sick of writing reports, sick of the FOB food–sick of everything. On the other hand, I was sad to leave the people we'd gotten to know so well. Some of them, like Ozzie, we'd helped move on to bigger and better things. But for most of the people–the Fi-Fis, Farquaads, and Old Man Friends–this was the last time we'd ever see each other, and I wasn't at all certain the future held anything good for them.

The platoon split up that last day, with each section making its way around the neighborhood to say goodbye to our various friends and acquaintances. My last stop of the day was, of course, at Dr. Manza's house. As I stood on the street in front of Dr. Manza's house, images of the hundreds of other times I'd stood in the exact same spot ran through my mind.

A dozen yards to my left was the crater where an IED had exploded when we were still battling for a foothold in the *mahala*. That IED had ruined a section of Dr. Manza's wall and blown out the windows in every house on his end of the street. A few months later, a stupid kid had thrown another IED into the same crater, but we'd caught him in the act and made him crawl into the hole to pull out the poorly made bomb himself.

A few yards to my right was the spot where we'd concocted a clumsy sting operation to arrest a two-man terrorist cell by using a hastily recruited double agent to ride in the bad guys' car and tip us off when they were close. The whole thing had worked surprisingly well, and we'd put the two bad guys in jail and gotten our guy out of trouble.

Looking around at the street, I marveled at how little had changed. Dr. Manza's gate was still the same white, slightly rusted decorative gate, the small white Daewoo that Mr. Faisal drove was still parked in the same place it had always been, and when he came out, Mr. Faisal still wore the same humble smile he had the first time I met him. He escorted me into Dr. Manza's house, gestured to the couch, and brought me a bowl of Jordanian almonds and a glass of tea. In a few minutes, Dr. Manza arrived. We hugged, greeted each other, and sat down.

Dr. Manza lit a cigarette, took a long drag, and gave me a thoughtful look.

"Do you remember the time we worked together to arrest the two bad men and recover the *qunbula* on my street? It was last summer, I believe."

"Yeah, I remember that, Doc. Why?"

"Do you remember asking me what I was thinking that day?"

I thought for a minute. "Yeah, I do remember that. You had a spaced-out look on your face like you were thinking about something, but you wouldn't say what it was."

"Yes, that was the day." Dr. Manza looked down at his teacup and put out his cigarette. "That day, I remember thinking about how sad it was that the things we would do to improve the *mahala* would not last very long after you Americans leave. I still think that."

"You know, Doc, I've been concerned about that, too. I think we can make it work though, if we can just get the next unit to keep things together here until the problems between the neighborhood and the government get resolved."

Dr. Manza looked up and gave me one of the saddest smiles I've ever seen. "No, that won't work either, Dan. It's us. It's the Iraqi people. We lived under Saddam for too long, and once you leave, we will go back to the old ways. If you stayed for ten more years, it would not matter. When you left, we would still go back to the old ways. We can't help it."

"Would that be so bad? Surely, most of the stuff we put together here will last. Look how happy everyone is. You don't think they'll work to keep it that way?"

"No, I do not think so. The new government is supported by Iran. It will take revenge on the Sunnis for the oppression of the Shi'a people under Saddam, and the Sunnis will fight back against the government. In our neighborhood, the people will hate the police again, and the police will hate us. The new security force will fight the police and either flee or be killed. This is certain."

"C'mon Doc, you don't think people can change? Surely there's hope?"

Dr. Manza laughed softly and shook his head. "Dan, you sound like that speech at the end of the old movie with the American boxer and the Russian boxer who fight one another."

I thought for a second, "You mean *Rocky IV*? The 'I can change, you can change' speech after Rocky beats Ivan Draco?"

"Yes, that one. It is an excellent movie. I like it because it is such an excellent portrayal of the American character. You Americans are optimistic, and you always think everything can be changed if you just work hard enough. That is wonderful, but it is also not true. Our ways in Iraq are very old, and they do not change like that."

"Really? You think it's completely impossible?"

Dr. Manza thought for a minute and laughed again. "If you want to change the people in Iraq, there is only one way you could do it."

"How's that?"

"Kill everyone here over the age of eight and start over with the next generation. Without the old generation in the way, the children can be fixed."

I laughed out loud. "Even you, Doc?"

"Yes, Dan. Even me."

22

Fort Riley, Kansas, May, 2008

The springtime weather the day we landed back at Fort Riley was hot, dry and, of course, windy. We'd landed about an hour before and boarded buses at the airfield to take us to the redeployment center, a small building on Custer Hill where the redeployment ceremony for the squadron would be held. There were fifty soldiers from across the squadron on the bus, and the mood on the fifteen-minute drive was a combination of excitement, exhaustion, and apprehension, as soldiers wrestled with jetlag after the long trip from Kuwait and mentally prepared for the upcoming reunion. When our bus arrived at the redeployment center, we were herded off, through a rear door in the building and into a small antechamber which adjoined a large stage. The stage was in an auditorium where the assembled families of the 1-4 CAV were waiting for their loved ones' return.

As we stood in the antechamber, listening to the excited burble of the three or four dozen unseen wives and children in the audience, the rear detachment NCO briefed us on how the ceremony was going to unfold.

"Alright, y'all. First, let me say welcome home. We'll try to get you out of here as quickly as possible. Here in a minute, the music is going to start. It's some good patriotic shit, Toby Keith, I think. When that happens, y'all will walk up those stairs in some kind of organized fashion and stand in formation on the stage. Your families are out there watching, so don't look like a bunch of dick bags. Once the song stops, the announcer will say some stuff, and at the end he'll cut you loose so you can run out into the audience."

He paused and checked his notes. "That make sense? Anyone got any questions?"

Nobody did, so we stood there sweating and fidgeting for a few minutes until a thunderous rendition of *Courtesy of the Red White and Blue* started playing on the building's audio system, and the soldiers of the 1st Squadron, 4th U.S. Cavalry walked up the stairs to catch a glimpse of the loved ones we hadn't seen in months.

"We" is a generalization, of course, since not everyone in the formation had family members in the crowd. Walters, for example, was a single guy. His plan was to get a ride from one of his buddies as soon as possible so he could go pick up his new F-350 and snag a case of beer. Several others in the formation had similar plans. For these guys, the ceremony was an unwelcome obstacle between them and a reunion with the good ol' U.S. of A. For most in the formation, though, the ceremony was an overly scripted, slightly irritating, but not *too* painful step toward reunion with the most important people in their lives.

Not for me, though. Alycia and I had agreed she shouldn't come to the ceremony. We'd participated in redeployment ceremonies before, and they rarely started on time or finished quickly. She was also seven months pregnant with the baby girl we'd conceived on R&R, so sitting on a bench to wait for a ceremony that would probably be delayed was not high on her list of pleasant activities, particularly when that ceremony also derailed Gareth's afternoon nap. Her absence made sense, but it also sapped any small amount of joy I might have gotten from the ceremony itself and made me impatient to get home. As I stood sweating in the back row of the formation, I tried to ignore my irritation and focus on what was going on around me.

Bright incandescent lights beat down on my head, and Toby Keith lyrics rang in my ears as I looked around the auditorium at the rows of deliriously happy families in the crowd–some weeping and others cheering. They all looked so happy. The soldiers around me looked similarly excited.[1] At that moment, in that room, it looked like all the problems that had transpired over the last year were forgotten. All the arguments, shouting, missed birthdays, suspicions, and tears lay in the past. They would come back of course–fifteen months' worth of wounds don't heal overnight, and they don't heal without some scarring–but today, all of that was drowned out by the overwhelming joy of being together again.

The song ended, the lights died down, and the announcer shouted, "Ladies and gentlemen, please join me in welcoming home the soldiers of the 1-4 Cav!" The auditorium erupted into applause and shouting. Soldiers broke ranks and ran toward the crowd,

---

1. Except the single guys of course. Since strippers couldn't get on post, the single guys had nobody tearfully waiting for them in the audience. They had the impatient, hungry look of guys looking forward to reunions of a different sort with the day-shift at the Foxy Lady Lounge in Junction City.

leaping off the stage in some cases and running down the stairs in others. Simultaneously, a swarm of wives and children ran toward them, embracing, or sometimes literally climbing, their returned soldiers. Even most of the single guys turned toward one another and gave each other bro hugs and back poundings. After fifteen months, we were finally home. Well, almost home anyway.

I stood watching for a minute, not really feeling like a member of either group, before walking back toward the door we'd entered to see if I could find a ride home from one of the rear detachment guys. Fortunately, as I was walking out, I bumped into an old friend–Matt Babiarz.

"Dan Pace! Welcome home motherfucker!"

"Holy shit, Matt! It's been a long time. I think the last time I saw you, you were on a spine board being carried out of sector when that grenade blew your ass up."

Matt laughed. "Yep, my career as a platoon leader was short-lived, indeed. I finally got out of that shitty ass headquarters XO job, and I got blown up on my third goddamn patrol. Have you ever seen a bunch of dudes sadder than the guys who had to carry my ass on that board, though? I bet I was pushing 350 pounds with all my kit on."

Matt had briefly been a platoon leader in Apache Troop. He had been leading a foot patrol in 840, the *mahala* just west of ours, when an insurgent had hurled a grenade at him that had exploded violently and shredded his leg. He'd been evacuated from theater and never returned. Now he was working for the rear detachment and helping bring people home.

"That was a hell of a thing, I'm sure. Didn't you break one of the spine boards?"

Matt laughed again. "Yep, I broke that shit right in half. It dumped me in the street in the middle of Baghdad. What a goat rodeo."

"That feels like a million years ago. How's the leg?"

"The Army fucked it up at first. I had to wait so long for surgery that one of the nerves in my knee was at risk of dying, which would have made me a cripple for life. I got Congress involved though and sorted that out, and now my knee is as good as new." Matt did a quick shimmy step to emphasize his point.

"That's awesome, man. Glad everything is going well. They going to keep you in the service?"

"Nope, the leg isn't *that* good. They're going to medically retire me."

"That sucks!"

Matt shrugged. "It is what it is. It's funny, though. I spent four years at West Point, then another two years in training before I finally got to combat, then I got my ass blown up in the first week on the ground."

I laughed. What else was there to do? "I'm sure you'll figure something else out. At least you don't have to go to any more XO meetings."

"Very true, man."

"Oh, hey, would you mind giving me a ride back to the house? I'm short a ride home."

"Absolutely. Your family still living on post?"

"Yep, down the hill in old post housing."

"No problem at all. Let me check in with Major Layborne, and we can roll out."

We walked around the building to find Layborne. He was standing near the front doors of the building, still welcoming people home. He was too busy for more than a quick greeting, so after Matt and I checked in with him, we walked across the parking lot to a cobalt blue Toyota FJ.

"You like the new ride?"

"She's beautiful, man."

I climbed into the passenger seat. Matt shifted the FJ into gear, rolled down the windows, and pulled out of the redeployment building parking lot.

As we rolled down the road, it struck me how unfamiliar all the familiar things felt. The first thing I noticed was the smell. The air in springtime Kansas has a hot, grassy aroma to it, which is similar to how the part of Central Texas where I grew up smells. It was a scent I hadn't specifically thought about for as long as I could remember. Today, though, the smell of grass, wind, and occasional grill smoke so sharply contrasted with the aroma of burning oil and rotting garbage that permeated Baghdad, that I couldn't stop drinking in the scent. As we drove past the 1-4 CAV's buildings, those converted garages and chow halls I'd spent so much time in prior to Iraq also struck me as a strange combination of the recognizable and the strange.

For example, at first glance, the Apache Troop headquarters still looked like the place where I'd so often walked in, flopped down behind my desk, and waited for Cook or Strong to shout at me about some random task. On closer inspection, though, there were details which had changed and rendered the place alien. The piles of cigarette butts outside the doors were missing, the guidon that indicated the commander's presence was gone, and the doors and windows were closed and locked. Changes that I couldn't see also flooded into my brain as we drove past. Long would never open the arms room again, Dixon would never ask me another question about Ranger school, and North and I would never go for another run together during morning PT. My memories of the headquarters were bound up with so many things that were now gone, and without those things, the building that remained was only a husk of the place I remembered.

We continued along Custer Hill, passing the old squadron and brigade headquarters buildings, both of which had become as foreign as the Apache Troop headquarters. We even passed the old 1-28 building, devoid now of its much-loathed lion. Each thing I saw initially appeared the way I remembered, but on closer inspection, turned out to be different in some critical way that rendered it unrecognizable.

Further on, we passed the post exchange, the base's equivalent of a mall, complete with dozens of stores, a huge parking lot, and a Burger King. The crowds of people, walking to and from their cars, eating at the tables outside, and corralling their children, made me openly gape. The blue jeans, shorts, sunglasses, and to-go bags looked so quintessentially middle-American that they shouldn't have caused me to even glance twice. I'd seen identical crowds a hundred-thousand times in my life. Today though, I found the typical American scene completely captivating.

Where were the stray dogs gnawing, humping, and scratching at fleas in the streets? Where were the mixed pools of motor oil and antifreeze glistening in the gutters? Where were the *hijabs* and *dishdashas*? Intellectually, I knew the scene I was looking at should contain none of those elements, but my endocrine system remained unconvinced, so it was producing feelings of wariness and uncertainty.

"You okay, Dan?"

I pulled myself away from the scene and smiled. "Yeah, Matt. I'm okay. I think it's going to take a minute to get used to being home again."

"Yep, I was the same way. It took a few weeks before I got back to normal."

The road continued to the end of Custer Hill and then curved down steeply as we wound down into the valley where most of the post housing was located. As we approached my house on Lower Brick Row, my stomach tightened. The moment I'd both been looking forward to and avoiding thinking about for months was about to be here. How would it go? Alycia and I hadn't talked much in months. It had gotten too hard to figure out how to fill a phone call–or to even want to make one-and I wasn't sure how that would affect our face-to-face reunion. I didn't want things to be awkward or distant. On the contrary, I wanted things to just be the way they had been before I had left.

"You sure you're okay, Dan?"

Matt must have noticed me spacing out again. "Yeah, I think so. I'm just trying to figure out how walking into my house is going to play out. I didn't spend enough time on the family this year, and I'm not sure what to expect when I get home."

Matt slowed the car, pulled onto the shoulder, and stopped. He looked over at me. "You think there's going to be trouble? She's not going to come at you with a knife or anything, right? We've had some domestic problems with guys, and I don't want to see you go down that path."

"Nah, it's not like that, Matt. We just haven't talked much lately, and our lives have gone in completely different directions. I'm not sure she even wants to see me. Shit, if I were her, I probably wouldn't."

Matt thought for a minute. He was a single guy, so the married dynamic was unfamiliar territory for him, but he'd been on rear-d for most of a year, and he'd certainly seen these sorts of problems play out frequently with other guys who'd rotated home.

"I don't know, Dan. How about when I drop you off, I'll wait around for a minute. If everything's okay, give me a thumbs up, and I'll take off. If something's not working, come back outside, and we'll head over to my place until things cool down."

"That works, man. Thanks."

Matt pulled his car back onto the road, and we turned onto Lower Brick Row and pulled up to 165 B.

My house looked exactly like it did when I left. Alycia's blue Camry was still parked in the driveway. As always, it was covered in sticky brown sap from the walnut tree whose branches overhung most of our yard and rained down thick goo this time of year. The yard was still freshly mowed, the clothesline in the same location it always had been, and the front door still decorated with a seasonal wreath of flowers. Everything was exactly as it had been.

Except it wasn't.

"Thanks, Matt. I appreciate the ride home, and it was good catching up with you."

"My pleasure, Dan. Give me that thumbs up when you're sure everything's good."

"I will, for sure."

I opened the car door, walked around to the FJ's back hatch, and pulled out my duffle bags. I carried them up the six concrete stairs to our door and tried the handle. It was locked. Our front door locked automatically when closed, and I realized I didn't carry a key to the house anymore. I reached for the doorbell, paused, and knocked instead. It was around naptime, and if Gareth was asleep, I didn't want to wake him.

Through the window inset into the door, I saw our cat, Mickey, walk into the kitchen in response to my knock and look at me. I wondered if he'd be alarmed; he was never happy to see strangers, but he seemed fine. He stood there for a few seconds twitching his tail, then walked off disinterestedly. "I guess he remembers me," I thought. That was something.

I stood there for a good twenty seconds, holding my bags and wondering if I should knock again. I looked back over at Matt, who was still patiently waiting. Should I borrow his phone to call? And then I heard the door latch click.

I looked back over, and there was Alycia, opening the door. She was wearing a sundress, and she was barefoot. Her long hair was loose, and she had a sad sort of smile on her face. She looked beautiful. We stood looking at each other for a few seconds, then she teared up a bit and said, "Hi. Welcome home. I'm a whale."

I looked down at her belly. She had gotten huge when she'd been carrying Gareth, and this time was no different. At seven months pregnant, she looked like a normal person trying to smuggle a beachball under her dress.

I didn't care at all. "You look amazing, and I'm really happy to see you."

I reached my arms out for her, and we hugged. In that hug, it felt like the ice that had been building up between us for so long melted almost immediately. All the emotion that had been pent up behind a wall of worry and neglect, the fear, the love, the longing, it all came flooding back, and we just stood there hugging in the entryway for a long while. Everything was going to be okay. It might not be the way it was before I left, but it was going to be alright.

With that hug, I realized Alycia was willing to look past my absence and neglect. We continued hugging for a minute at least, finally breaking off the embrace and looking at one another. I turned to give Matt a quick thumbs up and nodded to him. He returned the nod and drove off.

I turned back to Alycia and smiled. She returned the smile, tears in her eyes, and we walked hand-in-hand into the house and back into our life together.

23

Fort Riley, Kansas, May, 2009

I wish I could say that after that homecoming, everything returned to normal. A clean, happy reunion would have put a nice bow on the story of my deployment to Iraq, and it would have given *It Was What It Was* the sort of happily-ever-after ending that people enjoy reading.

Unfortunately, this wasn't how the deployment ended, so if I told the story that way, I'd be doing the people who lived through The Surge an injustice. Just as you can't really understand a book like *Band of Brothers* without watching *The Best Years of Our Lives*, you can't fully comprehend The Surge without understanding what happened to us after we came home. In contrast to World War Two, where most of the guys who deployed served one long tour, then came home and reintegrated–however painfully–into normal life, GWOT deployments weren't one-offs; there was no post-deployment period, there was only a series of intra-deployment periods.

For GWOT soldiers, post-deployment time became a refit period before our next trip back to one of the wars. Even single-term soldiers, who only served a four-year stint in the Army before returning to civilian life, often had two or more deployments under their belts, while for career soldiers, three, four, or ten deployments were normal during a twenty-plus-year careers. Consequently, wartime deployments weren't just one-time events to look back on and get over. Wartime deployments were the refrain in the song of our lives, and this endlessly repeated refrain defined and intruded on all aspects of our and our families' worlds.

By the time our unit had fully redeployed from Iraq, the 1-4 CAV's leadership already knew the date of its next rotation into theater. Fifteen months after coming home, the unit would return to Iraq again, and everyone in the unit knew it.

Over the next year, this impending return drove the unit, and the people serving in it, along a roller coaster refit that resembled the initial build-up in 2006. However, this time a few things differed. Two years ago, most soldiers in the unit were on their first deployment. Now, though, most of the NCOs who were preparing the unit for combat had been with the 1-4 CAV in Iraq. Also, most of the new leaders being rotated into the unit from elsewhere had at least one combat deployment under their belts. This changed the tone of the unit's preparations, for better and for worse.

On the one hand, the experience these leaders possessed meant that training was significantly more focused and realistic than it had been two years earlier. On the other hand, the cumulative wear and tear from repeat deployments—on both the soldiers and their families—caused new problems. Whereas my old friend Jeff, the guy I'd known from basic training who was pulled off of our last deployment for personal problems, was the exception previously, this time numerous guys were overwhelmed by personal issues. Our unit wasn't remarkable in this regard; this was an Army-wide phenomenon. From 2001 to 2007, the war improved the Army. Combat winnowed out poor leaders, bad equipment, and weak processes. It cleaned out the worst parts of the old Army and sharpened the best into a tremendously effective war machine. However, after 2007, the trend reversed. While equipment and processes continued to improve throughout the GWOT, soldiers and their families suffered increasing damage from repeated, long deployments, and this personal baggage started to outweigh the benefits of combat experience.

Alycia and I were no different. While it wasn't immediately apparent, the wear and tear of a year of neglecting our relationship had taken its toll. The first few months after I redeployed, it really seemed like everything was back to normal. Gareth and I bonded quickly, and our family enjoyed a honeymoon period during the summer of 2008 that was as good as any time we'd ever had. A completely relaxed work schedule facilitated this. As key figures like Crider, Jones, Callahan, Cook, and Hamilton rotated out of the unit, the 1-4 CAV lapsed into a vacation-heavy recovery period, when the soldiers worked very little and were at home most of the time. This overall reduction of stress on the system allowed our family the time it needed to start to heal.

But new leaders rotated into the unit that fall, and as the deployment timeline for our next rotation to Iraq came into focus, the unit's work schedule substantially picked up. Leaders like our new squadron commander, Scott Nelson, knew they had to prepare the unit for its next rotation, and they knew they had too little time to do it.

Unfortunately, as work picked up, I started spending too much time focusing on the mission again, and all the old wounds suffered in Iraq reopened. Before long, it was apparent that Alycia and I still had a long way to go to recover from my time away. Every little thing that happened seemed to trigger an argument, and it was hard to avoid bringing up old fights with little provocation.

Our family was hardly alone. Across the unit, as the training schedule picked back up, soldiers who were on the edge of personal problems started falling apart. Guys became alcoholics. Guys got divorced, divorced again, or were arrested for domestic violence. Guys stole cars. Guys did drugs. Guys fell apart professionally, putting on twenty or thirty pounds, failing fitness tests, and showing up late to work. One guy even put on his dress uniform to conduct a funeral ceremony, disappeared for a week, and was found by the police passed out drunk under a highway overpass–still in his uniform. And none of these were useless guys. They were the soldiers who had been our best performers in combat. But traumatic brain injuries, post-traumatic stress, and family problems took their toll, and for many guys, that toll became too high to pay.

In hindsight, as leaders in the unit, we didn't handle most of these cases very well. In 2008, many of us still didn't really accept that post-traumatic stress or the behavior changes traumatic brain injury can cause were legitimate problems. Instead, we often fell back on the old Army philosophy that bitching about your problems and allowing your personal issues to interfere with work was something weak people did; real dudes pushed through that shit. Consequently, when guys showed symptoms of PTSD or of TBI, we often chastised them or recommended disciplinary action instead of getting them the help they needed.

I wasn't any better at seeing these symptoms in myself at home either, and a good number of the problems my family and I had over the next year could probably have been alleviated–or at least better understood–if I'd had more perspective about the root cause of our issues and been able to do more early on to prevent them.

I finally hit a turning point in January when Nelson offered me command of Apache Troop. Kirk to Crider's Picard, Nelson and I got along well, and he wanted me to take the Troop to combat with the Squadron later that year. His offer was professionally flattering, and I seriously considered it, but I knew that if I took the Troop back to combat, I would once more be choosing work over my family. Thinking I finally had my priorities straight, I turned the offer down.

The trouble was, I wasn't sure what to do instead. I'd applied to the FBI, the Department of State, and I'd even looked into commercial opportunities, but in the depressed economy of 2008, there weren't many people knocking at my door with job offers-but the Army was. So many soldiers and officers were leaving the service that, for the first time in its history, the Army was offering officers $30,000 to continue to serve. Alycia and I talked about it, and we decided to stick with the Army, but to try out for the Special Forces to break up the back-to-back deployment cycle we'd been stuck in since 2001. I'd worked with Special Forces teams before, and I was interested in the mission. Also, joining the SF community would ensure Alycia and I had at least a few years together before I deployed again because the training pipeline was two years long.

After a few months of physical conditioning, I attended selection for the Special Forces regiment in January, and after a miserable couple of weeks, I was selected to start Green Beret training.[1]

The news that I'd passed the selection course, and the accompanying orders to leave Fort Riley, brightened our family's future considerably. It removed us from the 1-4 CAV's deployment cycle and gave us a fresh sense of hope that we'd have the time to work through our family issues. The news that I'd passed selection was less welcome at the 1-4 CAV. Nobody in the squadron was hostile, on the contrary, most people were congratulatory, but passing selection meant that I wasn't going to Iraq with the 1-4 CAV, and if I wasn't going to Iraq, I wasn't really part of the team anymore. Along with the guys who were injured, waiting for surgery, receiving counseling, awaiting disciplinary action, or leaving for some other reason, I was now an outsider. And as the year progressed, and the deployment team got tighter, the unit didn't have time for outsiders.

By May, I felt like a ghost haunting the house of a family I didn't know. Almost everyone else I'd worked with, from Rob Humphry and Rich Smith, to Matt Babiarz and 1SG Strong, were gone, and since I was working at the squadron headquarters now, I almost never saw the remaining guys from 1st Platoon. The weeks passed until my last day at the 1-4 CAV. After getting the final stamps on my clearing paperwork and turning in my last pieces of gear, I walked down to Comanche Troop to see if I could find the newly promoted Staff Sergeant McDowell to ask him how his deployment preparation was going.

When I arrived, he was sitting at the platoon computer, working through the details of a range he was running.

---

1. All of which was colorfully captured by the Discovery Channel in *Two Weeks in Hell*. I was roster number 151.

"*Staff* Sergeant McDowell. Holy shit, they'll promote anyone nowadays."

"*Captain* Pace. I guess they will! What are you doing down here?"

"This is my last day, bro. Thought I'd come by and see how you turds were running the place without me."

McDowell laughed. He'd had no problems dealing with redeployment, and his good natured sense of humor was as present as ever. "I think we're doing okay, sir. The new guys are dumb, like new guys always are, but I'm whipping them into shape. I heard you passed selection."

"I did, indeed. When are you going that way?"

"After this trip, sir. I can't wait."

"You got time for a beer later? I'd like to catch up before Alycia and I roll out. Any of the other guys who are still around are welcome to join us, of course."

McDowell looked up at the clock. "Shit, sir. Sorry, but no can do. We've got to run a range today, and we have an NTC rotation next week. I gotta brief the new commander on all the details in fifteen minutes, and I've got no time for anything."

There it was. The team was getting ready to leave, and there was no time left for outsiders. A year ago, my old friend Jeff had been the guy getting left behind, and I'd left him at Riley with barely a second thought. Now I was Jeff, and nobody had time for me either.

I nodded and smiled. "No issue at all, man. I get it. I'll see you around, McDowell. Keep your ass alive on this trip!"

McDowell laughed. "Shit, you know I will."

He stood up, and we bro hugged, and afterward, I walked out of the building to my car. As I drove by the converted chow hall that was still the squadron headquarters building, I saw the guidon outside the door that indicated the commander was present. Soldiers were bustling in and out of the building on important, deployment-related business. The building looked almost exactly like it had a year and a half ago.

However, it wasn't my headquarters anymore, and I didn't know any of the bustling soldiers. I drove down Custer Hill and met Alycia and the kids at our now-packed-up house. We loaded up our last few items, buckled the kids into their car seats, and drove away from 165 B Lower Brick Row.

As Alycia and I drove away, I hummed a line from the Army song, "...And the Army goes rolling along." The Army did indeed go rolling along, and now this part would roll along without me.

I took Alycia's hand and we smiled at each other as I guided our Camry through the Fort Riley gate and drove into the future.

Epilogue

Camp Vance, Afghanistan, 2019

"I have never advocated war except as means of peace, so seek peace, but prepare for war. Because war... War never changes."
Ulysses S. Grant

"Give me a kiss to build a dream on
And my imagination will thrive upon that kiss
Sweetheart, I ask no more than this
A kiss to build a dream on"

Satchmo's gravelly voice poured into my ears from the Bose headphones I was wearing to cut down on the background chatter in the JOC so I that could focus on the project I was working on. I looked up from the computer monitor in front of me and noticed the killcam feed on Screen Three. It showed the glowing white image of an old man and his donkey picking their way along a mountain trail. The outline of a bundle of RPGs, could be seen protruding from the donkey's saddle, strapped in place under a stack of blankets and a box of what were probably bootlegged DVDs.

I munched on a BBQ rib as I checked the grid on the feed. The old man was in Paktika province, not far from where I had been patrolling all those years ago when we'd been ambushed in the middle of a *Thompson Twins* song. Maybe this old man had been there, too. Was he one of those old men I'd seen slowly picking their way past us in the wake of that firefight?

"Shit, it's been almost twenty years. There's no way either of those two guys is still alive."

Nick turned in his chair and said something to me. I slipped one earphone off and muted Louis so I could hear him.

"What's up, Nick?"

"What was that you were saying, sir?"

"Oh. Nothing. Just reminiscing. What are we looking at on Screen Three?"

"The fires guy is calling it a Taliban facilitator and weapons trafficker moving equipment to prepare for an attack on GIROA forces. The guy is the trafficker, and the donkey is the facilitator."

I chewed on the rib and considered Nick's assessment. "Has JAG cleared it?"

"Yes, sir. JAG says it's good. Intel says it's good. And I've cleared the strike for collateral damage. No issues."

I finished the rib and wiped my hand with a paper towel. I took the headphones off and hung them on the rack at my desk, stood up, and walked over to Nick's desk.

"Hand me the target packet, and I'll get the boss to clear it."

While Nick prepped the folder, I looked around the JOC. The glare of the 32 screens that comprised the main wall of the room bathed the faces of the fifteen folks who worked for me in a bluish white light as they pecked away at their terminals. In contrast to the immensely bright screens, the overhead lighting was dim to the point of darkness, and the air in the room smelled almost aggressively sterile, as though the EMFs from the massive quantity of electronics in the room had killed every living thing in the air. The electronics seemed to kill the sound in the room too, and between the dampening hum they emitted and the fact that every employee wore heavy headphones that limited conversation, the JOC was eerily quiet.

"Yep," I thought, "everything looks about right."

Nick handed me the folder. I thanked him, took the packet, and headed to the commander's office. The walk took me down our memorial hall, in which hung a photo of every special operations soldier who had been killed in Afghanistan from 2001 to now. Green berets, red berets, tan berets. U.S. soldiers, British soldiers, and Canadian soldiers. Bearded, angry-looking guys in mismatched combat uniforms, and clean-cut guys in dress uniforms. Their pictures, over a hundred of them arranged from floor to ceiling, stared at me from the wall as I walked along it.

I reached the end of the hall and turned into the commander's office. Major General Buck Elton was seated behind his desk, reading glasses perched on his nose as he sifted through a pile of documents. He was in his late forties and possessed an almost photographic memory and a confident personality. He tolerated no bullshit, but he was also personable and engaging. As I walked in, he glanced up.

"Hey Dan. Got a strike for me?"

"Yes, sir. Guy and a donkey smuggling weapons."

"JAG good? Intel good?"

"JAG's good. Intel says he's nothing remarkable, but he's affiliated with the Taliban."

"Okay. Strike approved. Thanks."

I walked the packet back down the hallway, again passing under the eyes of the hundred-plus fallen operators, and returned to the JOC.

"Nick, the packet's approved. Make sure and tell Carly the strike is happening so she can work up a PR message on it."

"Gotcha, sir."

Nick turned back to his computer and put on his headphones. I climbed the short flight of stairs to my desk at the top of the JOC. The room was arranged like a movie theater, with me, Eric, and Jay occupying the highest seats so we could manage the entire floor.

I redonned my headphones and pushed play on the music, letting Louis Armstrong's voice once again fill my brain as I returned to work on the report.

"And when I'm alone with my fancies,

I'll be with you

Weaving romances, making believe they're true"

Out of the corner of my eye, I saw Screen Three flash brightly as the hellfire missile's detonation briefly washed out the killcam with white light. Aside from Louis Armstrong, there was no accompanying sound. I turned my head toward Screen Three, and in the explosion's aftermath I saw pieces of the dead man burning on the ground. The donkey, with the preternatural ability to hear incoming missiles so often demonstrated by his species, had bolted out of the blast area in time to survive, and was now grazing on whatever plant life grew from the rocks of Afghanistan, the RPG launcher still visibly protruding from his saddle.

The killcam continued to feature the grazing donkey as I looked back at my computer and continued typing on my report. A few moments passed. Then I saw Nick waving at me out of the corner of my eye. I slipped off one headphone and continued typing.

"What's up, Nick?"

"Sir, they want to reengage the donkey. They say the RPG counts as a weapon cache that needs to be destroyed."

"Nope. You know my policy, Nick. If the donkey survives the first hit, he gets to live."

To be honest, I wasn't the commander, and I didn't get to write policies. The strike guys knew that. If they really wanted to push the issue, they also knew they could get the general involved, and he might override me and approve the strike. But that would require them to get their commander to call the general, which would mean they would need to spin their commander up on the situation and convince him to care enough about a donkey with an RPG to bother calling his boss. Not surprisingly, they didn't push the issue, and after a few minutes, Screen Three panned off the donkey to find something else to stare at.

I continued typing my report, which was a compilation of assessment data on the Afghan special operations teams our guys trained. These teams had been working with us for almost two decades, and they were highly effective. Unfortunately, the rest of the Afghan Army was hot garbage, so the Afghan senior leadership ground the special operations units into oblivion by throwing them at every hard problem in the country. This meant the Afghan special operations guys often had a better relationship with us than with their own government.

As I pecked out my recommendations on the computer, I paused for a moment and sighed. We were making the same mistakes here that my platoon had made in the Surge in Iraq all those years ago. The only way to build a stable government in Afghanistan was by helping the Afghan government build its own institutions and earn the trust of its own people. However, we hadn't been able to do that. We'd instead built a bunch of workarounds that relied on us being there to make them function. If anything, the longer we'd been here, the *more* essential we'd made ourselves to the entire apparatus, rather than the opposite.

After almost twenty years, the U.S. military was still great at *fighting* the enemy. We'd even gotten good at *finding* individual bad guys, pretty much wherever they tried to hide. Unfortunately, neither of those things really mattered. Unless we wanted to keep killing the bad guys ourselves in perpetuity, we'd eventually need to work ourselves out of the job, but there was no way to do that. We didn't have the will, desire, ability, or time, to fundamentally change Afghan culture, and nobody was going to support Dr. Manza's recommendation to kill everyone over the age of eight and start over, so nothing we built was going to last any longer than we were willing to maintain it. Once we left, the tide would simply wash our sandcastle away.

As I chewed on my input to the report and another (now-cold) rib, I realized that in 2019 we were still doing basically the same thing my platoon had been doing in 2003-screwing with a bunch of people who just wanted everyone to leave them alone.

I saved the report, put back on my headphones, and leaned back in my chair to listen to the rest of one of my favorite songs. I stared blankly up at the wall of killcams, and my mind wandered far away to my house in Southern Florida. "I wonder what Alycia is up to today," I thought.

Oh, give me your lips for just a moment
And my imagination will make that moment live
Give me what you alone can give
A kiss to build a dream on

Afterward

To all the people I've forgot who never cross my mind
To all the good friends I have lost
I'm sure we've had good times
And as the time piece ticks away the seconds of our lives
We're left with only memories and glimmers of good times
Well, nothing ever lasts
Don't waste your life by living in the past
And raise your glasses to the sky, fuck yeah we had good times
                    — A Toast to the Good Times–Straight Outta Junior High

Fifteen years after The Surge, the last conversation I had with Dr. Manza still strikes me as one of the most profound moments of the deployment, and perhaps of my entire military career. His parting comment, that the only way to successfully transform Iraq into a western-style democracy was to kill everyone over the age of eight and start over from scratch, was meant in jest of course, but it pointed to a powerful truth about the Global War on Terror. Nation building, as we were attempting it in Iraq and Afghanistan, was impossible.

Throughout my generation's twenty years of constant wartime service, we attempted endless combinations of training, force structure, equipment, and strategy to find the magic solution that would allow our military efforts to create stable, just, and effective governments in Iraq and Afghanistan. Because of the history, culture, and regional situations of those countries, though, our efforts–*no matter how effectively implemented*–were doomed to fail. If Iraq ever overcomes the tension between the Sunni and Shi'a populations that leads to so much violence there or rids itself of the nepotism and corruption that plague its government, it will be because the Iraqi people, rather than the American military, want that to happen.

I came to that realization long after my time in The Surge was over. Since then, the rise of ISIS, the rise of the Muslim Brotherhood in Egypt, the partition of Libya, and the collapse of the Government of the Islamic Republic of Afghanistan have reinforced my belief in its truth. The funny thing, though, is that the realization hasn't tainted the mostly positive memories I have of my service in the GWOT.

In *The Myth of Sisyphus,* Albert Camus concludes that his protagonist—forever condemned to push a huge rock up a hill each morning, only to have it roll back down to the bottom each evening—was ultimately happy with his situation. Even though Sisyphus's task could literally *never* be accomplished, and was therefore purposeless, Camus believed Sisyphus derived satisfaction from the conduct of the task itself. I enjoyed my time in the GWOT for similar reasons. Even though there was no actual way to accomplish the tasks my fellow soldiers and I were given, most of us enjoyed the act of trying.

I can still vividly picture Sproul pouring his concrete pad on COP Banshee. I'm sure it didn't last long after we left. When we departed the COP and redeployed, no U.S. forces replaced us there, so within a year the pad was probably ruined. Whatever the Iraqis who own the concrete factory that now occupies that site have done with the place, the pad itself is most likely long gone. To me, though, the impermanence of that pad does not suggest Sproul shouldn't have poured it.

Pouring that pad was meaningful to Sproul. The act gave him a sense of purpose because he was solving an immediate problem and because the act itself was satisfying. Our efforts in the *mahala* were similar. Rebuilding curbs and cleaning up trash were satisfying and moving families back into the neighborhood to occupy abandoned houses felt good. Being thanked by mothers when their children attended school and being cheered by kids who were celebrating *Eid* were tangible indicators that we'd accomplished something positive. Even killing or arresting terrorists and recovering burning Iraqi policemen scratched our atavistic need to succeed and accomplish.

Fifteen years later, our efforts to rebuild *Mahala* 838, much like Sproul's work on the concrete pad, probably haven't amounted to much. The Fifis, Old Man Friends, and Dr. Manzas of the neighborhood have moved on with their lives and probably don't think about our efforts there very often—no more often than most of my former guys think about their time in Iraq. Many of them may not even remember our platoon as anything but one of several groups of helmeted, armored Americans who rotated through their lives during that tumultuous period. Similarly, the buildings we repainted and the playgrounds we built have probably been torn down, repurposed, or fallen into ruin, and our security force was certainly either disbanded or radicalized into the ISIS fighters we later helped the Turks kill in Syria. *Abu Tayara* has moved on, and the effects of our work there were either fleeting, or in some cases even counter to our desired outcomes.

For me though, even as all the tangible evidence of our work has likely vanished and our efforts have gone mostly awry, the sense of purpose and satisfaction I feel about what we accomplished remains. The 1-4 CAV deployed to Iraq, found a huge problem to solve, and for fifteen months, we tackled it with everything we had. When we left, we could put our hands on our hips and look at what we had accomplished with the satisfaction of a job well done.

Looking back at it all now, that sense of satisfaction helps me make sense not only of what we did, but what it cost us to do it.

About the Author

Daniel Pace grew up in Texas and joined the Army in 2001. He married his wife, Alycia, in 2001 and spent twenty-two years in the military, serving in Afghanistan, Iraq, South America, Central America, Africa, and Europe. Daniel retired from the Army as a Special Forces Lieutenant Colonel in 2023, and he and Alycia now live in northwest Florida with their five children.

It's hard for a book to get noticed these days. Please consider writing a review on Amazon to help me continue writing.

www.danielvpace.com

www.ingramcontent.com/pod-product-compliance
Lightning Source LLC
Chambersburg PA
CBHW060909120626
46553CB00001B/265